International Medical Device Clinical Investigations

A Practical Approach

Second Edition

Edited by
Herman Pieterse
Marja G. de Jong
Peter Duijst

CRC Press
Taylor & Francis Group
Boca Raton London New

CRC Press is an imprint of the
Taylor & Francis Group, an **informa**

T0203637

CRC Press
Taylor & Francis Group
6000 Broken Sound Parkway NW, Suite 300
Boca Raton, FL 33487-2742

First issued in paperback 2019

ISBN-13: 978-1-57491-085-8 (hbk)
ISBN-13: 978-0-367-39977-1 (pbk)

Library of Congress Cataloging-in-Publication Data

International medical device clinical investigations : a practical approach second edition /
    edited by Herman Pieterse, Marja G. de Jong, Peter Duijst
        p. cm.
    Includes bibliographical references and index.
    ISBN-10: 1-5749-1085-X (hardcover)
    1. Medical instruments and apparatus--Evaluation--Handbooks, manuals, etc. 2. Clinical Trials--Handbooks, manuals, etc. I. Pieterse, Herman. II. Jong, M.G. de. III. Duijst, Peter.
    [DNLM: 1. Equipment and Supplies--standards. 2. Clinical trials--methods. 3. Device Approval.
    W 26 I594 1997]

    R856.4.I58 1999
    610'.28'4--dc21

                                                                                          97-260071

Visit the Taylor & Francis Web site at
http://www.taylorandfrancis.com

and the CRC Press Web site at
http://www.crcpress.com

Authors/Editors

Herman Pieterse started his own consultancy and interim management company in 1991. As Managing Director and Principal Consultant of Profess® Medical Consultancy BV, in Heerhugowaard, The Netherlands, he now assists several major companies in the health care industry with regulatory and Quality Assurance issues, GCP clinical development, and training. He is a Certified Auditor for Quality Systems that conform to the requirements of BS EN ISO 9000 and BS 7229/ISO 10011 standards. Prior to the work with his consultancy and management company, he held various executive management positions at various leading pharmaceutical and medical device companies. Herman Pieterse studied pharmacology at the Free University in Amsterdam and has an M.Sc. degree in chemistry.

Marja de Jong has held executive management positions in the regulatory and quality assurance fields with leading medical device companies. In 1994 she founded her own consultancy and project management company, Medidas, in Wijk bij Duurstede, The Netherlands. Ms. de Jong is a member of the editorial board for GCPj and has published books and articles on numerous subjects in regulatory, quality, and clinical affairs. Ms. de Jong holds a masters degree.

Peter Duijst was appointed a Director of Aortech Europe Ltd., a Scottish cardiovascular medical device company, in May 1993 and a Director of the Company in January 1997. He is the Medical and Regulatory Affairs Director, responsible for the clinical and regulatory affairs of the Group. Prior to his appointment with Aortech, he held various executive management positions with various leading companies, which included senior research scientist at Shiley Regulatory Affairs International in The Netherlands, head of the Clinical Medical Research Bureau International in Arnhem, The Netherlands, and physician in the Department of Pediatrics, Academic Hospital in Amsterdam, The Netherlands. He has a masters degree in medicine, an M.D., and a Ph.D.

CONTRIBUTING AUTHORS

Philippe Auclair — Guidant S.A., Zaventem, Belgium

Willem Ezerman — Parexel/MIRAI BV, Amsterdam, The Netherlands

Jaap L. Laufer — EDMA, Eerbeek, The Netherlands

Kenneth R. Michael — KRM Associates, San Diego, CA, USA

Henriët E. Nienhuis — Parexel/MIRAI BV, Amsterdam, The Netherlands

Allison Oliva — Cardiovascular Dynamics Irvine, CA, USA

Patrick W. Serruys — Erasmus University, Rotterdam, The Netherlands

Robin N. Stephens — Ave UK Ltd., Burgess Hill, West Sussex, UK

Elisabeth A. M. van der Linden — Parexel/MIRAI BV, Amsterdam, The Netherlands

Elisabeth C. M. van der Velden — Deventer Hospitals, Deventer, The Netherlands

Contents

SECTION II: STRATEGY AND PROJECT PLANNING

4.   Device Classification and Risk Assessment   **95**

Marja G. de Jong and Peter Duijst

SECTION III: HOW TO PREPARE A CLINICAL TRIAL

SECTION IV: HOW TO MONITOR
A CLINICAL TRIAL

Preface

The demand for clinical evidence has become an increasingly important issue in the development of medical devices. This is reflected not only in the regulatory requirements but also by health care purchasers as health care reforms take place worldwide. A wider perspective is needed by manufacturers of devices as they face increasing global demands to devise safe, cost-effective devices and treatments in an increasingly cost-conscious health care environment.

The clinical investigation process for devices has taken on new forms in worldwide device legislation. Although present for years in the medicinal products industry, a practice which culminated in the International Conference of Harmonisation for Pharmaceutical Products' Tripartite GCP agreement, Good Clinical Practice (GCP) for devices is rather a new 'modus' of collecting clinical data. Manufacturers are faced with coping with systematic and disciplined methods of conducting clinical investigations. Planning and preparing the study diligently and carefully is pivotal. Device studies require a significant investment of the sponsor's resources and time. It is therefore important to ensure that design, conduct, and monitoring/auditing of device studies satisfy all requirements—GCP and regulatory.

GCP is the international ethical, legal, and scientific quality 'standard' for the design and conduct of clinical investigations which involve the participation of human subjects. Compliance with GCP requirements, defined in a wide-ranging set of documents, provides public assurance that the rights, safety, and well-being of subjects are protected at all times. Fundamental to the principles of GCP is The Declaration of Helsinki.

Conducting a study without proper planning is discouraged, as the regulations require valid results obtained in a systematic, disciplined manner. Conducting a study in accordance with GCP, using the elements defined in this publication, is to the benefit of all parties involved: investigators, subjects, authorities, and, most important, manufacturers who have obligations far beyond the mores, ethics, and legalities of GCP.

On this premise, the authors of this book intend to give the reader information from a practical perspective on how clinical investigations should be performed in accordance with GCP. Albeit many chapters reference the Medical Device Directive (MDD), the principles, philosophies, and methodologies explained here are equally applicable to Active Implantable Medical Devices (AIMD) and In-vitro Diagnostic (IVD) products. Although the book concentrates on the two major market sectors, the European Union and the United States, this is not to say that the principles expounded in this book are not applicable to other potential trial countries or areas, for example, Australia, Asia, Japan, and South America. Quite the contrary. The information given herein forms a 'recipe' book of *how to* plan, prepare, implement, and close out the clinical investigation, regardless of where in the world the trial site may be located.

The chapters in this book are relatively autonomous and may be read in any order. Chapter 1 defines the regulatory environment in the two major market places: the European Union and the United States. It defines in simple and practical terms the complexities of international GCP and the regulatory environment. In many areas the European Union's requirements are more complex than those in the United States, particularly in the areas of labelling, languages, and cultural nuances.

The section 'Why Undertake Clinical Trials?' looks at the philosophies of clinical trials, types of trials and how to manage mega-projects (e.g., development of design proposals, or clinical investigation project plans).

Good preparation of clinical investigation is a prerequisite for success. Much attention is paid to this element in the section 'How to Prepare a Clinical Trial.' Once the preliminary activities are defined, the implementation "How to Monitor a Clinical Trial" begins. The qualifications and training of monitors are essential to successful implementation. Information is given on major aspects of implementing the process, from monitoring to data management, clinic preparation, and adverse event reporting and handling.

Once all the data has been collected, a final or *clinical study report* is to be developed and the data is to be analysed. The final phase is the close-out and reporting of the data.

Sponsors and investigators faced with the challenge of implementing clinical investigations can use this book as a tool or 'how to' manual. These individuals will realise after reading this book that GCP is in effect nothing more than 'Good Communications Practices'.

NOTE: Every effort has been made to ensure that the contents of this book are factually correct. Because of the steady state of evolution in regulations worldwide, the reader is advised to consult with the appropriate regulatory authorities in countries where trials are planned. Neither the publisher nor the authors and editors accept liability for injury, damage, or losses arising from any materials published in this book.

Herman Pieterse
Marja G. de Jong
Peter Duijst
March 1999

The terms *clinical research manager, clinical research project manager, project manager, project leader, clinical research associate, project coordinator,* and *study manager* are treated as synonyms in this book.

1

Clinical Regulatory Environment

Marja G. de Jong

Medidas
Wijk bij Duurstede, The Netherlands

Peter Duijst

Aortech Europe Ltd.
Almere, The Netherlands

1: European Regulatory Environment

Amongst the laws which the European Union (EU) has been harmonising are regulations which apply to medical devices. The Active Implantable Medical Device Directive (AIMD 90/86/EEC) and the Medical Device Directive (MDD 93/42/EEC) as well as the In-vitro Diagnostics Directive (IVD 98/79/EEC) are primary pieces of legislation created under this harmonisation effort.

The directives define specific requirements that devices must fulfil if they are to be safe for use. The following paragraphs examine the main characteristics of the directives in regard to regulatory requirements and Good Clinical Practice (GCP).

REGULATORY ASPECTS

According to the directives, manufacturers are required to prepare technical documentation for each device. Such technical documentation must include not only technical aspects (e.g., pre-clinical testing performed, along with the results; materials; manufacturing processes; special processes; and clinical 'testing') but also evidence that the Essential Requirements (ERs) are fulfilled. A list of ERs is found in Table 1.1.

There are a number of activities to be undertaken by manufacturers to ensure compliance with the ERs.

- Determine whether the product in question and its principal intended use make the product a medical device as defined by the directive (MDD, AIMD, IVD). For example a medical device can be used for either

  - diagnosis, prevention, monitoring, treatment, or alleviation of disease;

  - compensation for an injury or handicap;

  - investigation, replacement, or modification of an anatomical part or a physiological process; or

  - anti-conception.

- Classify the product as defined by the directives. Classification in the European Union (EU) and in the United States have similarities—but also have differences. Table 1.2 gives a comparative overview of the classification system and lists some of the regulatory requirements pertaining to those classes as they are currently applied under prevailing device legislation in Europe and the United States. Characteristics for classification under both systems include, for example, the surgical invasiveness of the device and the duration of the contact of the device with the body.

Table 1.1. Overview of the Essential Requirements

General ERs

1. Devices must be safe; any risks must be acceptable in relation to the benefits offered by the device.

2. Devices must be designed in accordance with the latest state of the art technology. Risks should be eliminated (preferential), protected against, or warned against (least desirable).

3. Devices must perform in accordance with the manufacturer's claims and specifications.

4. Safety and performance must be maintained throughout the indicated lifetime of the devices.

5. Safety and performance of the devices must not be affected by reasonable conditions of transport and storage.

6. Any side effects must be acceptable in relation to the benefits offered.

Particular ERs

1. *Chemical, physical, biological properties*: This requirement involves extensive testing of the medical device in question. Testing can include, but is not limited to, biocompatability, toxicological, pharmacokinetics, pharmaco-toxicological, and physicochemical testing; materials testing and evaluation; finished device testing, including tensile strength and bond; etc.

2. *Infection and microbial contamination*: This covers sterilisation and manufacturing environmental controls, and packaging as applicable to sterile medical devices.

3. *Manufacturing and environmental properties pertaining to strength and stability testing:* Testing is dependent on type of device and use.

4. *Devices with a measuring function:* Such devices must demonstrate continued accuracy and stability.

5. *Protection against radiation:* This pertains particularly to devices which emit ionising radiation.

6. *Requirements for devices connected to or equipped with an energy source:* These requirements are related to electromedical products and include electromagnetic compatibility, power supplies, electronic systems, software/hardware systems driven by computer technology, alarms, the prevention of electrical failures, etc. Most of the aspects are covered by a wide range of IEC documents; for instance, IEC 601.1.

Table 1.1 continued on next page.

Table 1.1 continued from previous page.

---

7. *Information supplied by the manufacturer:* This defines specific labelling requirements, such as contraindications, indications, warnings/precautions, and intended use of the device, which must be included in the labelling. Sufficient information must be provided to allow for the safe and effective use of the device in accordance with its intended purpose. Any other uses for the device, other than intended by the manufacturer, become the liability of the user.

Clinical ER

1. *Clinical data:* This defines the need to have clinical evidence available for all devices.

Note: Specific additional ERs may apply for IVDs under 98/79/EEC.

| TIP | Since not all of the ERs are applicable to ALL devices, it is important to review the ERs listed in Annex I of the respective directive. The six General ERs (articles 1 to 6 of Annex I), as well as the one clinical data ER (article 14 of Annex I), apply to all medical devices! Of the remaining seven Particular ERs, it is necessary to determine which apply to the device in question, and to take into account the intended purpose of the device. |

- Establish a comprehensive *quality system* that covers either design and production (ISO 9001/EN 46001) or production alone (ISO 9003/EN 46003 in draft).

| TIP | The quality system management standards, for example, the ISO 9000 series and the EN 46000 series, are available from local national standards bodies. |

- Carry out a complete **risk analysis** of the devices based on a risk-benefit ratio. The evaluation should be performed in accordance with EN 1441: Risk Analysis for Medical Devices.

Risk analysis needs to be performed for all devices regardless of classification. Risk analysis includes a clinical evaluation to ensure that any undesirable side effects constitute an acceptable risk when weighed against the benefits of the device. Such an evaluation comes from **clinical data** or **evidence**. Clinical data, in some form or other, is an integral part of the overall technical documentation for CE marking (see chapter 4).

Table 1.2. Medical Device Classification

Class	European Union According to Directives	United States According to Case Report Form
I	• Low-risk product that is neither sterile or has a measuring function (e.g., bandage) • Subject to compliance with the ERs • Does not need to have intervention of a Notified Body for CE-marking	• Low risk • Subject to general controls which have been used prior to device amendments
I-sterile	• Low-risk sterile product which is to be used in sterile condition (e.g., syringes without needles) • Subject to the controls of the ERs • Requires involvement of a Notified Body due to the sterilisation and sterility processes/controls	
I-measuring	• Low-risk measuring product (e.g., thermometer) • Subject to the controls of the ERs in addition to any particular standards pertaining to measurement function • Requires intervention of a Notified Body for CE-marking	
II		• Subject to *performance* standards that have been developed or are under development • Need to meet general requirements in addition to specific requirements

Table 1.2 continued on next page.

Table 1.2 continued from previous page.

Class	European Union According to Directives	United States According to Case Report Form
II		• Commonly require a 510k application and clearance by the Food and Drug Administration (FDA)
II-a	• Medium-risk product (e.g., tubing intended for use with an infusion pump) • Subject to controls of the ERs • Requires intervention of a Notified Body for CE-marking	
II-b	• Medium-to high-risk product (e.g., reconstructive implant) • Subject to the controls of the ERs • Requires intervention of a Notified Body for CE-marking	
III	• High-risk/life sustaining product (e.g., heart valve, stent) • Subject to the controls of the ERs • Requires a design dossier as well as technical documentation • Requires Notified Body assessment of dossier and systems prior to CE-marking	• High-risk devices that are (a) life-sustaining; or (2) of substantial importance in preventing impairment of health; or (3) potential for causing high risk of injury or illness • Requires submission of a PMA • Requires *safety and efficacy* as well as performance data

- Carry out (pre-clinical) testing to obtain objective evidence that proves that the safety and performance can be maintained *over the anticipated lifetime of the product*. This involves a vast amount of technical, physical, chemical, biological, and mechanical testing, including accelerated age testing and in-vivo and in-vitro (animal) testing. The scope and the depth of the testing is dependent on the intended purpose of the device in question. For critical devices, such as long-term implants or life-sustaining implants, more pre-clinical testing will be involved than, for example, a syringe. Generic standards, which apply to most devices, as well as product-related standards, should be examined and used when and where applicable.

TIP	Standards have a quasi-obligatory nature. It is advisable to follow them as much as possible, where appropriate. In questions of liability, the legal test may be against the standards available.

- Undertake the collection of **clinical data (evidence)** to support the safety and performance of the device under normal clinical conditions of use.

- Establish and implement a **post-production feedback mechanism,** including vigilance reporting. This system encompasses many other elements of information processing and related activities, not just complaint handling and the reporting of adverse events.

Clinical Data and the Directives

Clinical data, the result of clinical 'testing', could be considered the independent assessment verifying a manufacture's claims when the tested device is used under (normal) clinical conditions. The primary purpose of clinical data is to provide **clinical evidence** to demonstrate that the device, when used correctly

- performs as intended,

- is safe and effective in its use,

- complies with the ERs, and

- has neither undesirable side effects nor produces adverse events that are unacceptable when weighed against the benefits to the patient

The directives provide some clues as to what clinical data entails:

1. The directives, in Annex X (directive 93/42/EEC) and Annex VIII (directive 90/86/EEC), define clinical data to 'includ[e] the collection of literature and/or the result of clinical investigation.' Hereby two mechanisms are given by which clinical data may be gathered: *evaluation and clinical literature* (a.k.a, clinical evaluation) and *clinical investigation.*

2. Clinical data in particular are required in cases of devices falling within Class III and implantable and long-term invasive devices falling within Class II-a or II-b.

Hence, clinical data may be collected through one of the following methods:

- a review of existing data, medical and non-medical;

- a clinical investigation (i.e., clinical trial); or

- a combination of both.

Clinical Evaluation

Clinical evaluation, is in principle, an evaluation or appraisal of *available data* related to the device (or similar versions/specimens) and concerning the safety and performance of the device in clinical use. Evaluation encompasses the use of a variety of sources which give an *historical and technical perspective* on the device in clinical use. The most common sources of information include

- non-clinical, non-medical, or technical data (e.g., risk analysis, design and production failure modes, design and manufacturing processes);

- clinical or medical data (e.g., review of final reports of clinical studies undertaken);

- post-marketing surveillance and trend data, including review of adverse incidents and market performance; and

- relevant scientific literature.

 Performing a literature search in medical databases (e.g., MEDLOG, MEDLINE, and EMBASE) is feasible. Such database searches help to define relevant and applicable literature.

Post-marketing surveillance data is a vital source of information, in particular for existing, established technologies and devises. For devices which have been on the market for many years, ample historical technical data (both medical and non-medical) should be available to demonstrate compliance with the ERs.

Some of the information available, however, may be questionable from a scientific basis, and could render the data inadequate for demonstrating compliance with the ERs. In such cases, it may be necessary to acquire a medical expert opinion be a clinical investigation.

If an expert opinion is needed, it should come from a clinician who is familiar with the device and its technology. This could be an individual eminent in his or her field. It is important that the expert's opinion be an unbiased view about the actual use and performance of the device in a clinical setting.

TIP To gain acceptance of the expert opinion's report or assessment of the product by regulatory authorities, documenting the qualifications of the expert per his/her curriculum vitae is recommended.

TIP When in doubt about the scientific validity of existing data, obtain a medical expert opinion.

Clinical Investigation

Clinical investigation, on the other hand, refers to any systematic study of human subjects undertaken to verify the *safety and performance* of a specific medical device under normal conditions of use. It is a means of verifying the anticipated behaviour of the device in situations where there is no other means of demonstrating device behaviour (i.e., safety and performance) through the use of available data. The need for clinical investigation should be a fundamental consideration of the design and development process of a device. A clinical evaluation (i.e., a review of available data) should always be undertaken as part of the design process to aide in determining whether or not an investigation is necessary. When a clinical investigation is to be undertaken, it must be conducted in accordance with the terms of GCP.

Regulatory authorities[1] are taking the position that device trials are needed in the following circumstances to satisfactorily demonstrate compliance with the ERs:

1. a completely new device is proposed for the market, whose components, features, and/or method of action are previously unknown;

2. an existing device is modified and the modification might affect clinical safety and performance;

3. a previously established device is proposed for a new modification; or

4. a device incorporates new materials coming into contact with the human body or existing materials applied in a location not previously exposed to that material, and for which no convincing prior chemical experience exists.

A review of published scientific literature, for the collection of medical or clinical data concerning the safety and performance of devices similar to the new (unproven) device in question, should, nonetheless, be done. An examination of publications concerning devices that are 'substantially equivalent'

---

1. NB-MED/ 4.2.4 and 4.2.4a/4.2.4b.3 December 1995. 'Recommendations on Clinical Evidence and Guidence on Clinical Trials.'

(to use a U.S. Food and Drug Administration [FDA] phrase) to the device in question may help a company determine what must be done to prove the safety and performance of the new (unproven) device.

| TIP | When it is necessary to substantiate the claims being made and determine safety and performance (and in some cases efficacy), clinical investigations are performed. |

Other factors also influence the necessity for undertaking clinical investigations[2] including, for example:

- device (medical) economy (i.e., less costly, but equally effective, as other devices);

- therapeutic modality, whereby efficacy is influenced by the operator of the device;

- institutional differences;

- improvement of quality of care;

- reimbursement issues.

When What Applies

Manufacturers are faced with taking into consideration many factors (regulatory and non-regulatory alike) whilst determining the need for clinical investigations. Discussing specific circumstances and devices with the regulatory authorities is advisable.

| TIP | Always discuss your clinical data gathering plan with the Notified Body (NB) prior to beginning the research. Should there be additions to the clinical data, it is best to understand and be aware of these in advance rather than at the time of certification or a surveillance audit. |

An overview of when to apply the two processes (clinical evaluation or clinical trial) is given in Table 1.3.

2. Atchia, H. 1997. 'Ethical aspects of clinical evaluation and investigation'. *RAJ* .5 (40) (November): 280–290.

It is evident from Table 1.3 that in most cases clinical investigation is *not* likely to take place. Clinical evidence obtained by appraisal of available data could be the most frequent mechanism used. However, other factors influence the determination of the necessity of undertaking clinical investigations, implying that a wider perspective concerning clinical investigations is indeed necessary.

TIP	Whether evaluating available data for clinical data/evidence purposes, or undertaking a clinical investigation from scratch, all data used must have been collected in a scientifically valid manner. Should this not be the case, it is advisable to obtain a medical expert opinion and/or an NB opinion of the data in a critical and unbiased manner.

Table 1.3. What Is to Be Done When

Clinical Evaluation	Clinical Investigation
• Used in cases of existing products, materials, indications of use, and/or technologies which have an established and proven track record • Information is obtainable from many sources of *available* data including in-house data at the manufacturer's site and/or scientific publications relevant to the device in clinical use • Scientific validity of the available data needs to be determined. If data is or could be scientifically questionable, then a clinical investigation may be necessary, or, an expert written opinion may be sufficient.	• Conducted for regulatory reasons, in cases of all Class III products and implantables (Class II-a and Class II-b) • Conducted on device when its behaviours (safety [efficacy] and performance) cannot be satisfactorily or adequately demonstrated by other means • Could be conducted for non-regulatory reasons (e.g., medical economy, improvement of quality of care, examination of therapeutic modalities) • Must be performed in accordance with GCP • Should always include, as appropriate, an appraisal of available data (including literature) as part of the data collection.

Use of Foreign Data

Data used in foreign submissions *could* be of use in meeting European clinical and technical data requirements. However, it is important to carefully review and analyse the clinical data to ensure it is scientifically valid, complies with the European requirements, and is relevant to the device in question—particularly to its safety.

Manufacturers are well advised to discuss any foreign clinical data with the appropriate authorities in order to ensure the data is sufficient to fulfil the requirements of the ERs.

GOOD CLINICAL PRACTICES (GCPs)

The provisions in the directives define GCP obligations that device manufacturers must comply with at all times and in all situations when clinical investigation is undertaken in the European Union.

GCP is an international ethical and scientific quality requirement for designing and conducting clinical investigations that involve participation of human subjects and set of Quality Assurance management tools that provides public assurance that the rights, safety, and well-being of the trial subjects are protected, consistent with principles originating in the Declaration of Helsinki, and that trial data is correctly and properly recorded and credible.

GCP Under the Directives

The following ethical and legal provisions comprise device GCP as required by the directives and further amplified by the harmonised standard EN 540, Clinical Investigations of Medical Devices for Human Subjects' and its international counterpart, ISO 14155, "Clinical Investigations of Medical Devices":

- Clinical investigations must have as main objectives the assessment of safety of the product under normal conditions of use, and that the benefits outweigh the risks.

- Clinical investigations must take into account all ethical considerations as defined in the Declaration of Helsinki and its amendments, so that every step in the clinical investigational process is regulated—from the first consideration of the need and justification of the study to the publication of the results.

- The manufacturer or his/her authorised representative must adhere to GCP at all times.

- Clinical studies must be conducted using the latest scientific methodologies and technical knowledge.

- A *clinical investigational plan* (or *protocol*) is to be used.

- All medical procedures used must be appropriate to the device under investigation.

- The study must be performed under normal conditions of use.

- The device under investigation must have defined features of safety and performance.

- The study must have a risk-benefit proportionality so that there are no undue risks for the subjects.

- Data concerning the device must be provided in advance to the investigator(s).

- A statement (or document) must be drafted by the manufacturer which includes

  - identification of the device;

  - a clinical investigational plan;

  - an Ethics Committee (EC) opinion;

  - the names of institutions and practitioners participating in the study;

  - a statement declaring that the device is in conformance with the ERs (apart from those aspects covered by the clinical investigation) and that all precautions have been taken to protect the subjects' health and safety.

This document could be the *investigator's brochure (IB)*.

- Studies conducted under the directives (i.e., for devices in Class III and implantable or long-term invasive devices in Class II-a or Class II-b) require submission of a clinical investigation application to the CAs of the countries where the clinical investigation sites are located, along with a 60-day waiting period.

- Favourable EC approval is required prior to commencing any clinical investigation.

- A written final report analysing the results and presenting the conclusions of the investigation, *signed by the investigators.*

- Appropriate recording, handling, and reporting of AEU arising during the clinical trial to the appropriate authorities and institutions.

- Appropriate labelling of trial devices, including the phrase 'for clinical investigational purposes'.

- Use only qualified and trained investigators and monitors.

- Use of informed consent and other relevant information needed to properly inform all subjects about the clinical investigation and the consequences, risks, and benefits.

Regulatory Notification

Under the directives, each CA in the countries where the clinical investigational sites are located needs to be notified of the clinical investigation prior to commencing the trial. In practice, this means that a clinical investigation application may have to be submitted to the CAs as required. Most countries may have pre-defined formats and/or requirements for such applications. Copies of these pre-defined forms should be obtained directly from the appropriate CAs. It is not advisable to file an application using a self-made format. Table 1.4 gives an overview of requirements country by country, illustrating the national nuances as well as differences existing amongst countres in the European Union.

| TIP | Obtain any clinical trial application forms from the CAs. Avoid the use of homemade versions of the forms. |

Although national nuances do exist, the common elements contained in these application forms and appended data include

- the clinical trial application, completed, signed, and dated as required by the CA(s) of the country(ies) where the trial sites are located;

- toxicological and biocompatibility data to ensure product safety;

- the clinical trial plan;

- copies of informed consent to be used, preferably in the language of the country in which the trial site is located;

- the Investigator's Brochure;

- Ethics Committee approval from the clinics where the trial is to take place or a copy of the application to the Ethics Committee;

- the labelling provided with the device (this may be required to be multilingual);

- the curriculum vitae (CV; résumé) and/or competence certificate of the clinical investigators;

- information about the device and how it is intended to be used (optional for some types of device trials);

- the ERs checklist or statement that the product meets the ERs of the applicable directive; and

- insurance coverage information for the subjects (provide a copy of the policy).

Table 1.4. Overview of Clinical Investigation Requirements in the European Union*

Country	Observations	Labelling	Subject Information	Insurance	Ethics Committee (EC)	Notification
Austria	• Reimbursement of device, in some cases, regulated by individual's insurance/health care providers • Detailed clinical trial requirements defined in the law • Auditing of clinical trial is mandatory • Following GCP is mandatory	• German	• German	• Required by law • Personal injury insurance applicable • No minimal limit for damage compensation; usual amounts range from 5 million to 7 million Austrian shillings per patient. • Ensure for coverage by Austrian insurer.	• Mandatory for all clinical trials • Regional EC in most area of Austria; some also have hospital EC areas • Multi-centre trials require one EC approval, although all ECs must receive list of participating investigators • Hospital • May charge fees	• Mandatory notification • For trials conducted under the directives, a 60-day waiting period is obligatory • No waiting period for other trials • No fees • No standard form required
Belgium	• No reimbursement for clinical investigation devices	• French, Dutch/Flemish, or German • English may be permitted	• French, Dutch/Flemish, or German	• Required by law	• Hospital • Some may charge fees • Applicable to all clinical trials	• Notification for all trials, but should be checked • Authorisation may be given prior to 60-day period if favourable EC opinion is received for a clinical trial conducted under the directives • May start immediately after favourable EC opinion for all other trials • No CA fees • No standard form at this time

*The information represented in this table is based on current national regulations available. Since there is a steady state of flux in the regulatory arena in Europe, the reader is advised to consult with the appropriate authorities in advance of conducting trials in Europe.

Table 1.4 continued on next page.

Table 1.4 continued from previous page.

Country	Observations	Labelling	Subject info	Insurance	Ethics Committee (EC)	Notification
Denmark	• No reimbursement for devices	• May be Danish or other language • Must be agreed to by the EC	• May be Danish but must be agreed to by the EC	• The law does not mandate the necessity of insurance • CA considering to requiring insurance by law • Insurance recommended in order to meet GCP	• Central Scientific Ethical Committee, part of the Danish Ministry of Health (i.e., CA) • Must submit for approval for all types of clinical trials • Fees charged	• Favourable approval by EC does not require notification of CA • For all clinical trials, no waiting period once favourable opinion is given
Finland	• Drafting clinical trial guidelines • Reimbursement dependent on hospital/patient insurer	• English may be permitted if labelling for professionals	• Finnish or Swedish if intended for patient	• Law does not mandate necessity of insurance • Insurance recommended to comply with GCP	• Hospital • Fees exist • Applicable to all clinical trials	• Notification required for all devices • 60-day waiting period for clinical investigation under directives • Fees for CA notification based on level of risk associated with product. The higher the risk, the higher the fee. • All other trials may start directly after notification to Ministry of Health, provided there is a favourable EC opinion • Standard form may be applicable

Country		Language	Insurance	Ethics Committee	Notification
France	• No reimbursement for clinical investigation devices • Loi Huriet applies to clinical investigation (i.e. French GCP) • MCCI require one principle investigator • Devices to be given free-of-charge to clinical investigation sites	• French	• Insurance (specifically, liability and risk) required by law • Insurance with a French company is mandatory • Duration may be set by EC for certain products	• Regional • Charges fees, approximately FF10,000 • Applicable to all clinical trials	• Must notify Ministry of Health for all trials • No delay in starting once a positive EC is granted • No standard form to be submitted, but should consult with Ministry of Health in advance due to changes within the Agency currently taking place • No fees
Germany	• Lander authorities supervise the clinical investigation process. Each Land could impose additional requirements. • May be dealing with 16 different sets of requirements at Land level • No reimbursement for clinical investigation devices/treatments	• German or another Community language, according to MPG • Lander authorities could require only German be acceptable	• Insurance required by law (Probandenversicherung) • Coverage set for a minimum of 1 million DM for death or incapacity to work	• Registered with and approved by CA • Currently 22 EC have appointed and recognised under MPG • Private institutions • Fees applicable • Applicable to all clinical trials	• Notification for all trials is required by the Lander authority where clinical investigation site is located • Technically, a 60-day waiting period is required if trial is conducted under applicable directives. However, Lander authorities may regulate this differently; trial may commence upon favourable Ethics Committee opinion. • Use of standard form may be required, depending on the Lander authority • No fees charged

Table 1.4 continued on next page.

Table 1.4 continued from previous page.

Country	Observations	Labelling	Subject info	Insurance	Ethics Committee (EC)	Notification
Greece	• No reimbursement for clinical investigation devices/treatments	• Greek, including software	• Greek	• Required by law	• Hospital • Applicable to all clinical trials • Charging of fees— unknown	• Notification required for all trials involving products which are not CE-marked • No notification required for trials using CE-marked products • Application may be in English, French, or German, although English is preferred • Must wait 60 days prior to beginning the clinical investigation for those trials conducted under the directives • No fees known
Iceland	• Devices may not necessarily be reimbursed. This must be checked with the Competent Authority.	• CAs expect English or a Nordic language	• Icelandic may be necessary, although this is not regulated by law. CAs expect English or a Nordic language.	• Law does not mandate necessity of insurance • Insurance recommended in order to meet GCP	• Hospital • Fees may be charged by the ECs • Applicable to all clinical trials	• Notification required for trials conducted under directives • Notification may be necessary for trials not conducted under the directives. Requires checking with CA. • 60-day waiting period for clinical investigations conducted under the directives • Considering charging fees • No standard form

| Ireland | No reimbursement for clinical investigational devices/treatments | English | English | Insurance necessary. Authorities must be satisfied that investigators and clinical sites are indemnified in case of injury. | Hospital; May charge fees; Applicable to all clinical trials | Notification on all clinical trials required; Class I and other devices used in trials not conducted for directives purposes, may commence clinical trial immediately after notification, provided a favourable EC opinion has been issued; For Class II-a, II-b, and III devices, notification to DOH at least 60 days prior of making devices available for clinical trial; Within 60 days of giving notice, DOH gives written notice to sponsor if objections to trial; there are otherwise, trial may commence if favourable EC opinion is obtained; No fees; Standard form may be under development |

Table 1.4 continued on next page.

Table 1.4 continued from previous page.

Country	Observations	Labelling	Subject info	Insurance	Ethics Committee (EC)	Notification
Italy	• Reimbursement depends on hospital and insurer • Clinical trials must be done in public hospitals only • Must follow Italy's GCP laws	• Italian, including all labelling on or provided with the device, as well as any software	• Italian	• Required by law	• Hospital • May charge fees • Applicable to all clinical trials per decree 4 6 24 February 97.	• Notification required for all clinical trials • Application may be in Italian, English, or French, although the Competent Authority prefers Italian or English • Must wait 60 days for clinical trials conducted under the directives. EC approval may run in parallel to 60-day waiting period. • Other trials require notification but can commence directly upon favourable EC opinion • No fees at present

Leichtenstein	• No specific requirements for clinical trials, although GCP must be complied with • No reimbursement for clinical trial devices/treatments.	• German	• German	• Required	• Fee structure not determined, although some EC may charge fees • Hospital • Applicable to all clinical trials	• Notification for all trials done under the directives • 60-day waiting period for trials under the directives • Other trials may commence directly upon receipt of favourable EC opinion • Notifications may be done in English, French, or German • Fee structure not determined at present • No standard format
Luxembourg	• Trial devices can be expected to be free-of-charge to clinical investigation sites	• French, German, or Letzeburgis • English may be accepted for professional use	• French, German, or Letzeburgis	• Required	• Hospital • May charge fees • Applicable to all clinical trials	• Notification for all clinical investigations which are being conducted under the directives • Applications may be in French, German, or Letzeburgish. • 60-day waiting period for all clinical investigations conducted under the directives • Other clinical investigations may commence directly after a favourable EC opinion is issued. • No fee structure determined as yet • No standard form/application

Table 1.4 continued on next page.

Table 1.4 continued from previous page.

Country	Observations	Labelling	Subject info	Insurance	Ethics Committee (EC)	Notification
Netherlands	• CE-marked devices may be used in clinical investigation if the device is being used in accordance with its intended purpose • Reimbursement depends on the hospital and insurer • Changes to the clinical trials requirements are anticipated and could be effective by mid-1998	• Dutch, although English may be accepted. Requires approval from CA • Clinical investigation devices must include a statement that they cannot be CE-marked	• Dutch	• Risk insurance required	• Hospital, but also Central EC • May charge fees • Applicable to all clinical trials	• Specific guidance issued by health authority, outlining the procedure for clinical investigation application and data to be presented to CA • Applications may be in English • Must wait 60 days for clinical trials conducted under the directives • Trials may commence directly after favourable EC opinion for CE-marked devices used in clinical investigation has been obtained • No standard form, although there is a requirement for specific documentation to be submitted as part of the application • No fees

Country						
Norway	• No reimbursement	• English or Norwegian	• Norwegian	• Liability insurance is the responsibility of the institution involved in the clinical trial • Insurance for sponsor not yet regulated by law • Insurance recommended to meet GCP	• Hospital • May charge fees • Applicable to all clinical trials	• CA must be notified of clinical investigations for regulatory purposes • 60 days given to review and comment on the application • May not commence trial until written permission is obtained from CA • Notification of other clinical investigation may be necessary. Checking with the authorities is recommended. • Fees not charged • Clinical investigation guidance being worked on by the CA
Portugal		• Portugese	• Portugese	• Required	• Hospital • No fees at present, although this could change	• Notification for all clinical trials is required • 60-day waiting period for clinical trials conducted under the directives • Other clinical trials may commence directly after favourable EC opinion is obtained • Applications may be submitted in Portuguese, English, French, or Spanish • No fees charged by CA at present, although this could change • No standard forms at present

Table 1.4 continued on next page.

Table 1.4 continued from previous page.

Country	Observations	Labelling	Subject info	Insurance	Ethics Committee (EC)	Notification
Spain	• A number of additional requirements exist in Spain which do not exist elsewhere • Clinical investigation is covered by a separate decree • Autonomous regions are entitled to require details of all clinical trials being conducted in their territories • Devices to be provided free-of-charge to hospitals	• Spanish	• Spanish	• Required by law	• Mandatory under law • Hospital • Fees may be charged	• Notification for all clinical trials is required • 60-day waiting period for clinical trials conducted under directives • Other clinical trials may commence 15 days from receipt of acknowledgement from Ministry of Health; favourable EC opinion required • Recommended to verify with MOH • Application to be in Spanish • Must notify Ministry of Health of any modifications to protocol or trial overall • Final report of clinical trials to be provided to Competent Authority • No fees at present • Standard format for application required

Sweden	• National Board of Health regulates the clinical trials • Medical Products Agency is responsible for assessing trial design and performance • Swedish guidelines exist, delineating the adherence of clinical trials to GCP • No reimbursement for non-CE-marked devices • Reimbursement given for CE-marked devices used in a clinical investigation	• Swedish, although English may be acceptable	• Swedish	• Required	• Hospital • Fees may be charged	• Notification of all clinical trials to authorities • 60-day waiting period for clinical trials using non-CE-marked devices • Applications should be in Swedish, although English will be accepted • Applications to be done in four-fold • Fees charged (Skr 10,000) • Standard format to be used. Applications must be signed by investigator, sponsor, monitor, and chief of the department at the clinical investigation site
Switzerland	• Cantons retain authority to licence hospitals for clinical investigation. Hence, advisable to use licenced hospitals. • Reimbursement is dependent upon canton where clinical investigation is being conducted, and are also dependent on contract between hospital and sponsor	• German, French, or Italian • Software may be in English	• German, French, or Italian	• Mandatory	• Hospital • Fees charged	• Notify the Federal Office of Public Healthcare before they begin any clinical investigation • Application may be in English • Standard application form • Fee charged if application not properly completed. Fee increases with each deficiency found.

Table 1.4 continued on next page.

Table 1.4 continued from previous page.

Country	Observations	Labelling	Subject info	Insurance	Ethics Committee (EC)	Notification
United Kingdom		• English	• English	• Regional for each trial site • Fees charged, and are variable	• No statutory insurance requirements. • Insurance recommended in order to comply with GCP.	• Notification is required for all clinical trials which are being done under the directives • Must give at least 60 day prior notice to MDA of trial • MDA has 60 days to object. If no objections are given, trial can commence if favourable EC opinion is obtained • May submit application to CA for multicentre sites when a favourable EC has been obtained for one of the sites. Other EC's opinions, however, must be submitted to CA prior to commencing trial in those hospitals. • Applications to be submitted in English • Standard application format mandatory • Fees charged, linked to risk classification of device

Manufacturers must wait 60 days before commencing a trial, unless national regulations for devices state otherwise (see Table 1.4). This does not mean that a manufacturer must submit the application to the authorities 60 days in advance. The clock starts when the CAs receive the application and appended data in their offices. A determination of the time which they need for review will depend on the type of technology or product involved. For an unproven technology or product, the review time may be more than 60 days. In such cases, the CAs could inform the sponsoring company that more time is required for the review. However, should a positive Ethics Committee approval be included with the application, additional, extensive review by a CA may not be necessary. A positive Ethics Committee approval should be an adequate incentive for the CAs to not respond and let the sponsoring party commence the trial. However, this may differ from country to country.

TIP	It is helpful, when new technologies are involved, to undertake a formal presentation to the authorities, prior to presenting the clinical trial application.

Marketing trials, which involve a clinician's subjective opinion and do not involve recruitment of subjects may not necessarily have to be done in accordance with GCP. However, once the objective of the marketing trial becomes one of proving safety, performance, and/or efficacy and uses human subjects, then it is no longer a marketing trial!

TIP	When there is a question of proving safety, performance, and/or efficacy, GCP requirements, are applicable!

Notified Bodies and Competent Authorities

It is the NB's responsibility to review any and all clinical data used as clinical evidence for proving the product's compliance with the ERs. An NB's review should concentrate on assessing whether the data included in the technical documentation and/or design file are sufficient, whether all of the elements have been covered as defined in the directives, and whether the NB's members are satisfied that the objective evidence is in accordance with the requirements. Therefore, it is feasible to discuss with an NB any clinical evidence plan on which a company may embark, be it a clinical evaluation or a clinical investigation.

TIP	Meet with the selected NB, well in advance of a Conformity Assessment procedure, to discuss the clinical data plan. Explain the plan and The rationale supporting your decision. If the device technology is new, present the technology in detail to the assessors.

A CA's responsibilities are to review the application, to ensure that no trials are taking place which have not undergone Ethics Committee approval, and to ensure that the method in which the data are being gathered is scientifically valid.

As previously stated, an overview of the current requirements for clinical investigations in the European Union is provided in Table 1.4. A list of addresses and telephone numbers for information pertaining to clinical trials is given in Table 1.5.

PRIVACY/DATA PROTECTION

Apart from compliance with the above mentioned, other requirements such as the way data protection and privacy of subjects are guaranteed, need to be taken into consideration by manufacturers contemplating any form of device trial in Europe.

Under European Directive 95/46/EEC 'Concerning the Protection of Individuals with Regard to Processing of Personal Data and On the Movement of That Data,' there is a prohibition of processing health care data unless there is explicit and unambiguous consent from the subjects for use, processing and handling of the data, either directly or indirectly. This directive is supplied by various national requirements in each Member State concerning the doctor-patient relationship and the privacy governing data and information exchanged and/or provided during the course of that relationship. An overview

Table 1.5. List of Contacts for Authorities Dealing with Clinical Trial Applications

Austria

Federal Ministry of Labour,
  Health, and Social Affairs
Stubenring 1
A-1010 Vienna
tel: +43 1 711 72 42 06
fax: +43 1 711 72 41 24

Belgium

Ministry of Social Affairs,
  Health, and Environment
Cite administrative de l'État
Quartier Vesale
B-1010 Brussels
tel: +32 2 210 48 90
fax: +32 2 210 48 80

Denmark

National Board of Health
Pharmacotherapeutic Division
Frederikssundsvej 378
DK-2700 Copenhagen
tel: +45 44 88 91 11
fax: +45 42 84 70 77

Finland

Medical Devices Centre
National Agency for Medicines
Siltasaarenkatu 18A
P.O. Box 278
FIN-00531 Helsinki
tel: +358 9 3967 22 50
fax: +358 9 3967 24 24

France

Agency du Médicament
  Direction des Laboratories et
  des Contrôles
Unité "Contrôle Nationale de
  Qualite/Enregistrement con-
  troles des reactifs"
143-147 blvd. Anatole France
F93285 Saint Denis Ledex

tel: +33 1 48 13 24 01
    +33 1 48 13 23 78
fax: +33 1 48 13 23 56

Germany

Bundesministerium für
  Gesundheit
Am Probsthof 78a
D-53121 Bonn
tel: +49 228 941 1170
fax: +49 228 941 4919

BFARM[1]
Medizinprod.
Seestrasse 10, Berlin
Germany
tel: 49 30 45 48 53 85
fax: 49 30 45 48 53 00

Greece

Biomedical Division
Ministry of Health, Welfare, and
  Social Security
Aristophoneous Street 17
10187 Athens
tel: +30 1 524 0649
fax: +30 1 524 7067

Iceland

Medical Device Division
Ministry of Health and Social
  Service
Laugavegur 116
IS-150 Reykjavik
tel: +354 560 9700
fax: +354 551 9165

Ireland

DoH
Hawkins House
Dublin 2
tel: +353 1 67 14 711
fax: +353 1 67 11 947

[1]BFARM is moving to Bonn in the course of 1999.

Table 1.5 continued on next page.

Table 1.5 continued from previous page.

Italy

Divison two
Department of Health
 Professions
Ministry of Health
Piazza P. le Industria 20
I-00144 Rome
tel: +39 6 5994 2423
fax: +39 6 5994 2111

Liechtenstein

Amt für Lebensmittelkontrolle
Im Reitacker 4/Postfach 402
9494 Schaan
tel: +41 75 236 7320
fax: +41 75 236 7310

Luxembourg

Ministere de la Santé Curative
Division de la Medicine
Rue Auguste Lumiere 4
L-1959 Luxembourg
tel: +352 47 85 634
fax: +352 48 03 24

The Netherlands

Ministry of Health
WVC
P.O. Box 20350
2500 E F Den Haag
tel: +31 70 340 79 11
fax: +31 70 340 78 34

Norway

National Board of Health
PO Box 8128 Dep.
N-0032 Oslo 1
tel: +22 24 89 53
fax: +22 24 90 17

Portugal

for active medical devices:
INSA
Avenida Padre Cruz
P-1699 Lisboa
tel: +351 1 757 35 57
fax: +351 1 757 36 71

for *non active medical devices*

INFARMED
Parque de Saúde de Lisboa
Avenue do Brasil nº 53
P-1700 Lisboa
tel: 351 1 790 85 00
fax: 351 1 795 91 16

Spain

Subdireccion General de Produc-
 tos Sanitarios
Ministerio de Sanidad y Consumo
Paseo del Prado 18/20
E-28071 Madrid
tel: +34 1 596 4347
fax: +34 1 596 4400

Switzerland

Medical Device Division
Swiss Federal Office of Public
 Health
CH-3003 Bern
tel: +41 31 322 9803
fax: +41 31 322 7646

United Kingdom

MDA
Hannibal House
Elephant & Castle
London SE 1 6TQ
tel: +44 171 972 8123
fax: +44 171 972 8111

provided during the course of that relationship. An overview is found in Table 1.6.

These requirements have wide implications, affecting all data gathered in the European Union, and any data *coming into* the European Union. Data must be controlled wherever it is processed. For the health care sector, these laws basically have given legal impetus to the Hippocratic oath 'I shall not disclose any personal information concerning my patients to any one else'. Hereby, data and records for clinical investigation must

- *not* contain sensitive data;

- *not* be linked back to the subject;

- be anonymised;

- be lawfully obtained;

- contain no patient data of a personal nature, such as birth dates, initials of subjects, or identification number that can be linked back to the subjects,

- all data used in a clinical investigation must be explicitly approved (i.e., through informed consent). Explicit approval from subject for use of all data in clinical investigation must be obtained.

- personal data shall not be permitted to leave the hospital, (i.e., go to the manufacturer or to any third party).

 Ensure one is familiar with the privacy laws in the Member State where the clinical trial sites are located. With the divergence of national data protection laws in the offing, each Member State will be treating data protection and privacy differently.

Other Related GCP Elements

Device clinical investigations conducted for purposes other than those defined by the directives are also obligated to be conducted in accordance with GCP. Table 1.7 defines the GCP documents which apply in such cases.

Table 1.6. Supplementary GCP Requirements

Year	Item	Key Elements
1997	European Directive 95/46/EEC the protection of individuals with regard to processing of personal data and on the movement of that data.	• Horizontal directive with far reaching implications • Prohibits processing of health data unless there is explicit and unambiguous consent from the subjects for use, processing, and handling of data directly or indirectly • Prohibits the transfer of personal data from hospital to any third party without explicit consent from the subject; • Data must be anonymised at all times to ensure that patient identity is continually protected • If data is to be processed by third parties, subjects must be informed with guarantees that data will be safeguarded, and anonymised • Subjects have the right to object to the use and handling of data by third parties; • All data processed must be safeguarded by a third party, (who will restrict access and maintain confidentiality).
	National requirements on clinical practice, conduct of medical practices, and data protection and privacy	• Each Member State has national requirements governing the providing of health care and the patient-doctor relationship. As part of those requirements provisions are incorporated to ensure that clinical investigations are always done within the confines of GCP, data is protected and privacy ensured; confidentiality ensured; and transfer of personal data and so forth. • Violations of those requirements have ramifications for the participating investigators, clinics, hospitals, and sponsors.

Table 1.7. GCP for Device Trials Other than those Required by the Directives

Year	Item	Key Elements
1996	ISO 14155 (Clinical investigation of medical devices)	• International standard applying to clinical investigations of medical devices for which clinical performance needs assessment • Defines the methodology and the documentation required • Defines the requirements for the conduct of clinical investigations;
1997	Council of Europe Convention of Human Rights and Biomedicine	• Primary purpose is the protection of individuals (right's of subjects and their privacy) • Applies to all clinical investigations undertaken in the health care sector regardless of products or objectives • Stresses that all clinical investigations are to be done in accordance with GCP • Sets forth the requirements to be abided by to ensure subjects are protected at all times

The Council of Europe's Convention on Human Rights and Biomedicine makes no distinction between clinical trials conducted on approved or non-approved products; on pharmaceuticals or devices; on biologicals or biotechnology products; nor between trials conducted for regulatory, research, commercial, post-marketing, economic, or other purposes. *Any* study which involves the use of human subjects must always be conducted to protect human subjects.

The philosophies and principles expounded by the Convention are identical to those mentioned under the section GCP and the directives. For a better understanding of the GCP principles and philosophies, readers are well advised to examine the International Conference on Harmonisation (ICH) Tripartite Agreement on GCP and its collateral guidelines concerning GCP. Although strictly applicable to medicinal products, its philosophies and beliefs give extra dimension in understanding GCP in general and, to how GCP is to be applied in practice to device trials in particular. Abiding by the spirit of GCP is important, regardless of the type of trial being conducted (e.g., medical device or in vitro diagnostic) or the type of product being investigated (e.g., pharmaceutical or biotechnological).

What Applies When

In summary, anytime human subjects are involved in a study, the study must be conducted in accordance with GCP as defined by the appropriate legislation. Table 1.8 gives an overview of the aspects to be taken into consideration when initiating a clinical investigation in the European Union.

Summary of Legal and GCP Aspects

Slowly but steadily the European Union is moving clinical investigations of all types of health care products into the realm of bio-law, which is enforceable and where sanctions can be imposed on all parties involved in cases of non-compliance. It is this element that is critical to the device manufacturer. The legal requirements pertaining to the conduct of clinical investigations are

Table 1.8. Clinical Investigation Overview and Criteria

Clinical Investigation	Criteria to be Fulfilled
Clinical investigations required by the directives (e.g., active implantable products, Class III devices, implantables, and long-term invasive devices in Class II-a and Class II-b) and for those circumstances on page 12:	• be conducted in accordance with GCP as defined in the directives and amplified by EN 540; • have a clinical investigation application submitted to the CAs in the Member States where the trial sites are located; • not commence until after the waiting period of 60 days unless national laws deem otherwise (see Appendix to this chapter); • have favourable EC opinion; • must take into consideration all elements of privacy and data protection as defined by the European directive, as well as any national requirements; • take into consideration any national nuances which may exist in the Member States where the trial sites are located.
Clinical investigations conducted on devices (other than those previously mentioned) where data is being used to demonstrate compliance with the ERs:	• do not, in principle, require the submission of a clinical investigation application to the CAs in the countries where the trial sites are located; • are required to have a favourable EC opinion before commencing the investigation; • may commence immediately after the handing down of the favourable EC opinion;

Table 1.8 continued on next page.

Table 1.8 continued from previous page.

Clinical Investigation	Criteria to be Fulfilled
	• must be conducted in accordance with GCP as stipulated in the directives and amplified by EN 540;
	• must take into consideration all elements of privacy and data protection as defined by the European directive, as well as any national requirements;
	• must take into consideration any national nuances which may exist in the Member States where the trial sites are located.
Clinical investigations conducted for other purposes (e.g., scientific, economic, post-marketing):	• do not, in principle, require the submission of a clinical investigation application to the CAs in the countries where the trial sites are located;
	• are required to have a favourable EC opinion before commencing the investigation;
	• may commence immediately after the handing down of the favourable EC opinion;
	• must be conducted in accordance with GCP (e.g., Declaration of Helsinki, Convention and ISO 14155/EN 540);
	• must take into consideration all elements of privacy and data protection as defined by the European directive as well as any national requirements;
	• must take into consideration any national nuances which may exist in the Member States where the trial sites are located.

always applicable where the collection of data and the protection of the subjects' rights are concerned. It is no longer sufficient to focus on only the device directive and the commercial puposes of a company; a broader perspective applies as additional laws, regulations, and guidelines come into effect.

The clinical investigation requirements stipulated in the directives apply in principle not only to clinical investigations required for Class III devices and implantable and long-term invasive devices in Class II-a or Class II-b, but also to any clinical investigation which a manufacturer may decide to initiate, even if it is not required under the Directives *per se*. This implies that any clinical investigation being conducted using any device other than, for example, active-implantables, Class III, Class II-a, or Class II-b implantables, or long-term invasive products, is also to be conducted in accordance with the requirements stipulated in the directives.

The following tips should be taken into account when commencing upon the investigation of a device:

TIP   Pre-clinical testing cannot be considered part of the clinical testing or clinical investigation *per se*. The results from pre-clinical testing are valuable in determining the safety profile of the device. However, the results may be extrapolated to clinical situations only within limitations, and with restrictions. It does not obviate the need to undertake clinical assessment, through literature searches or clinical investigation.

TIP   Ensure that in the developmental phase of the device under investigation all elements are taken into consideration and reviewed to determine whether clinical evidence can be gathered from available data or, alternatively, must be obtained through clinical investigation.

TIP   Exhaust non-clinical routes or options before making a decision to conduct a clinical investigation. Manufacturers are encouraged to discuss options and issues with the appropriate European regulatory authorities.

2: The U.S. Clinical Investigation Requirements Under the IDE

Under the FDA requirements, clinical studies are governed by the Investigational Device Exemption (IDE) requirements, section 520(g) of the Food, Drug, and Cosmetic (FD&C) Act, amplified and updated by the Safe Medical Devices Act. The purpose of the IDE process is to encourage the discovery and development of useful medical devices[3]. With the changes occurring in the FDA requirements, the IDE process has become less straightforward. With the changes in the IDE process, determining what data are needed for pre-market approval has become more difficult. Clear guidance from the FDA is required.

TIP	Ascertain early on what data are required for device approval. Pre-IDE meetings are important.

The IDE requirements are part of the overall pre-market approval (PMA) and/or 510(k) process for the market clearance of a medical device. In practice, an IDE is permission to undertake a clinical investigation to gather data concerning product safety, efficacy, and performance. Due to the long review time for IDEs, from six months to one year could be added to a product's approval time.

Once it has been ascertained what data are required, data collection can be commenced. After correlation and analysis, the result is clinical evidence for the submission of a market clearance application for the product in question.

The objective of the second part of the chapter is to review the process of completing clinical trials for FDA approval. Clinical trial requirements are specified in the IDE documents.

---

3. Kahn, J. 1993. 'Navigating the USE IDE process.' *Clinica* 553:10–12.

GENERAL

The USA medical device regulations[4] include the following provisions:

- validated scientific evidence necessary to support the effectiveness of a device (this may be done through well-controlled studies or partially controlled studies);

- studies and objective trials without matched controls;

- well-documented case histories; and

- reports of significant human experience with a marketed device.

The IDE regulation specifies which investigations are subject to the IDE requirement and which are not. The IDE requirement defines all of the principles of GCP as defined in the European medical device legislation (Annex X of the MDD and Annex VII of the Active Implantable Medical Device Directive AIMD) as amplified by EN 540. The IDE, however, is more restrictive than European legislation. It defines in great detail the exact content, layout, and format of the IDE application. In Europe, standard applications are necessary in some cases, but the final layout of the submission is not defined *per se.*

The IDE document defines

- requirements for investigation of non-significant risk and significant risk devices;

- how to format an IDE application;

- requirements for the protection of human subjects and Ethics Committees (ECs) or Institutional Review Boards (IRBs); and

- responsibilities of the sponsor, monitor, and investigator.

---

4. 21 Code of Federal Regulations, Part 860

IDE PROCESS[5]

An IDE may be defined as applicable for any device which may be subjected to a clinical investigations. Clinical investigations cannot be performed in the United States without appropriate IDE submission and approval (Figure 1.1).

Devices that are subject to the IDE requirements are handled according to the significant risk associated with the device. A significant risk device is one associated with high risk of the product. Examples of significant risk products include

- implants;

- life-sustaining devices; and

- devices used in the diagnosis, treatment, or prevention of disease or impairment.

Alternatively, devices that do not pose a serious risk to human health are considered to be non-significant risk devices.

Figure 1.1. Clinical Investigations

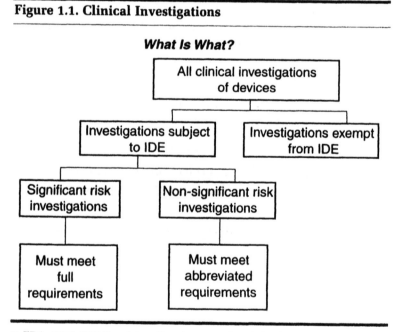

5. FDA. 1989. *IDE Regulatory Requirements for Medical Devices.* Rockville, MD: Food and Drug Adminstration.

The determination of significant versus non-significant risk is the decision of the sponsor and the IRB. The decision is based on the device's total risks—acceptable and unacceptable risks.

| TIP | The initial risk determination is made by the sponsor. The use of risk assessment can be helpful in determining whether a device is a significant risk or non-significant risk device under the IDE requirements. |

Hence, it is advisable to undertake a Failure Mode and Effects Analysis (FMEA) or Fault Tree Analysis (FTA). Table 1.9 provides an overview of non-significant and significant risk. The final decision about significant versus non-significant risk

Table 1.9. Overview of the Requirements for IDE

Non-Significant Risk	Significant Risk
The device is to be labelled with the name and address of the manufacturer, quantity, contra-indications/indications, hazards, adverse effects, interference with other devices or substances, and the statement 'Caution: Investigational Device. Limited by U.S. law to investigational use.'	The device is to be labelled with the name and address of the manufacturer, quantity, contra-indications/indications, hazards, adverse effects, interference with other devices or substances, and the statement 'Caution: Investigational Device. Limited by U.S. law to investigational use.'
Informed consent	IRB approval
Monitor investigational sites.	FDA approval
Record maintenance and retention for both sponsor and investigator.	Informed consent
Clinical trial reports for both sponsor and investigator	Monitor investigational sites.
Audit sites.	Record maintenance and retention for both sponsor and investigator.
Do not promote devices, test market them, or commercialise them in any way.	Clinical trial reports for both sponsor and investigator
Do not unduly prolong the investigation; strictly adhere to the plan.	Audit sites.
	Do not promote devices, test market them, or commercialise them in any way.
	Do not unduly prolong the investigation; strictly adhere to the plan.

is, however, the decision of the IRB. Their decision is influential on whether a submission is to be made to the FDA.

| TIP | Applications for significant risk studies are subject to FDA submission and approval. CAVEAT: Should the IRB determine that a device is a non-significant risk, while originally it was determined by the sponsor to be a significant risk, the FDA must be informed. |

| TIP | Working closely with the IRB ensures mutual understanding of the device and its rationale. Presentation to the IRB about the device, and its acceptable and unacceptable risks, is advisable. |

Significant Risk Studies—FDA Approval

Preparing and submitting an application to the FDA can be a difficult task to accomplish. The content of the application includes the

- name and address of the sponsor;
- complete reports of any prior investigation;
- complete investigational plan or summary;
- description of methods, facilities, and controls used for production, processing, storage, and installation;
- investigator's agreements (may be a sample);
- list of investigators with addresses and names;
- evidence that the agreements have been signed either by a statement or a matrix overview listing;
- list of the IRB members and their addresses;
- complete identification of the clinical sites where the study is to be done;
- price charged for the product, if any;

- justification why sale of the device does not constitute commercialisation;

- list of any claims pertaining to environmental impact legislation prevailing in the United States;

- labelling of the device;

- informed consents to be used; and

- information to be provided to clinical subjects (patients).

This is not a complete, comprehensive listing; other items may be requested depending on the device being submitted.

The FDA requires IDE supplements to be submitted whenever there is a change made in the clinical investigational plan that can affect the scientific soundness of the device or the rights, safety, and welfare of clinical subjects. What these changes are is constantly changing. It appears that the FDA may also intend to require IDE supplements for even insignificant manufacturing changes made to the device during the clinical trial. It is therefore important to discuss with the FDA what is required. However, guidance may not always be available on which changes fall within the purview of regulatory requirements.

> **TIP** Basically, any changes to the device or the clinical investigational plan that may affect the form, fit, and function of the device require notification and approval from the FDA.

Unlike the European application format, the U.S. requirements define particular layout and sizing requirements. These are to be followed laboriously.

> **TIP** U.S. submissions are to be on U.S. format pages (8.5 × 11 in.) paper, with 3-hole punch, multiple copies (3), and a left margin of 3.5 cm (approximately 1.5 in.). Applications exceeding 5 cm in thickness must be split into volumes, each numbered accordingly (e.g., Vol. 1 of [total volumes], copy 1).

The IDE study for significant risk devices may not be commenced unless approval from the FDA is forthcoming. The FDA will provide written approval for commencing the study even if there is a positive IRB approval. This is different from the requirements in Europe: A positive Ethics Committee approval may not necessarily result in a written response from the Competent Authority (CA) to commence the trial. There is, however, a 60-day waiting period in most European countries for trials done under the directives. This difference should be kept in mind when undertaking trials in the United States or Europe.

SUMMARY/COMPARISON

Although there are significant cultural differences between the United States and Europe, there are certain common elements that affect all clinical investigations. The following elements should be considered.

- All pre-clinical testing is to be completed prior to planning a trial.

- The clinical investigational plan or protocol and all relevant documentation pertaining to the clinical study are to be prepared and followed. Cultural nuances need to be taken into consideration.

- IRB/Ethics Committee approval is necessary prior to notification to regulatory authorities, if required.

- The trial is not to start before the appropriate written approvals have been given.

- Trial site and documentation are required to be audited/inspected.

- All processes must be validated before commencing a clinical investigation.

- The IRB/ECs and where required, the regulatory agency are to be notified of any modifications to the clinical investigational plan or protocal, device, or data that could affect the form, fit, function, and scientific validity of the trial and data.

A study sponsor needs to establish early on what the clinical design is to be and what is needed for the regulatory approval of the device, where necessary. Misunderstandings, assumptions, and serial disapprovals of clinical investigation applications may result if the issues are not defined up front. This is of critical importance with the FDA. The FDA's expectations appear to be a continuously moving target.

The following hints should be noted when doing clinical investigations in the United States:

- start an early dialogue with the FDA;

- determine scientific and validated endpoints at the beginning of the clinical design; and

- define what changes require supplements.

| TIP | It is better to overinform than to underinform. |

In the clinical trial atmosphere, U.S. companies need to understand that transferring a clinical study to Europe may not always be a favourable option, particularly when the data are to be used for U.S. regulatory submissions. The design of the trial is crucial to acceptance by the FDA when it reviews PMA and 510(k) applications. The focus in such reviews is mostly on the details of the device—its impact and its labelling; bias in the trial will be examined; and the objectivity of measurements is analysed. Where quality of life is involved, a validated tool is necessary.

It is the objective of trials to provide safety and effectiveness data about the device and to establish the benefit and real risks of the device in clinical use. Clinical data, in other words, is meant to ensure the clinical relevance of the device.

Harmonisation of clinical investigations done in the United States and in Europe for regulatory purposes is difficult due to the wide variety of differences. One of the many differences in regulatory trials is that U.S. trials are required to prove efficacy; European trials, performance. However, efficacy is becoming a major element in reimbursement trials.

In doing international clinical trials, the challenge for any manufacturer is to integrate the requirements into a clinical platform document. Differences in treatment, diagnosis, classification, and conditions of disease exist. Cultural differences are to be taken into consideration. For example, surgery for profit may be the objective of some American institutions, whilst in Europe, the practice appears to be disease management. Medical specialisms may differ from country to country, and even region to region.

The design of the investigational plan or protocol should also be taken into consideration. Questions that should be asked include the following.

- How are the control devices treated and/or imported—locally or nationally?

- Are local management techniques, when used as controls, significant to the FDA? To European regulatory agencies? To Ethics Committees?

- How far can an investigator go with data interpretation?

The last point is of importance since there is wider autonomy for European investigators in data interpretation where there are agreements made beforehand about who will do data analysis.

In order to define a harmonised global clinical investigation ensure that the design of the trial is simple, with end points that are easily measured objectively and are clinically relevant. It is important not to clutter the study with other non-definable end points. Multilingual case report forms (CRFs) may be feasible; for instance, English and the language of the country where the study is being done. Some companies may, however, not consider this feasible due to the added costs involved. But to ensure accurate and valid data collection, as many trials are done in regions where English is not the primary language of the medical specialists, this may be an option worth considering.

Finally, more than anything else, an examination of the pharmaceutical experiences in clinical investigations is important. Too often, medical device companies run clinical investigations as if they are marketing studies. With the movement toward global, harmonised requirements, the grey line between pharmaceutical-type and medical device-type investigations is beginning to fade. Evidence is already emerging with certain specific medical device types that the authorities are drawing on pharmaceutical. Medical device manufacturers are cautioned that if they think doing a clinical trial in Europe is considered an easy task that can be done by the marketing departments, they could be facing a rude awakening.

Section I

Why Undertake Clinical Trials?

2

Design in Clinical Trials

Peter Duijst

Aortech Europe Ltd.
Almere, The Netherlands

The various phases in the development of a medical device focus to a large extent, if not entirely, on the basic issues of whether the device performs according to its expectations and whether the device may be used on human beings. Once necessary laboratory testing has been successfully completed, and an acceptable performance and safety profile has been demonstrated, preparations may commence for the final, crucial step toward the general clinical environment (i.e., a study will be set up to test the device under clinical conditions).

Clinical testing will be necessary not only for newly designed devices. It may also be necessary for an existing device, although previously tested and approved for use under specified clinical conditions, when an alternative application has been developed but not clinically tested.

A clinical trial is a powerful tool; it structures the process and targets the acquisition of applicable patient data regarding a particular device. It further reveals a well-planned experimental set up, which intends to evaluate the clinical application of the device, and which intends to assess effectively the safety to an individual patient. The clinical conditions under which a trial is performed—therapeutic, prophylactic, or diagnostic—define the type of trial; the structural and organisational aspects will determine which trial design is utilised.

CLINICAL STUDY

Before discussing the various types and designs of clinical studies, it is important to have a good understanding of what is generally meant by the term *clinical study*. Various definitions exist, usually focussing on the application of drugs, rather than devices. Whether device or drug, it is essential to realise that we are dealing with an investigation that has been *planned according to scientific principles and is applied to human beings*.

In a well-chosen number of subjects, the performance and safety characteristics of the device are determined, in such a way that the outcome may also be extrapolated to the general population. This differs from reports on isolated, individual, clinical incidents (case reports). Although case reports represent a definite merit to the clinic, they also feed the suggestive idea that clinical relevance goes beyond the isolated event described and has a far-reaching importance. No matter how well case reports on the application of a device may have been worked out and documented, they cannot make a reasonable estimate of what the results will be once the device is clinically applied on a broader scale.

| NOTE | Device testing in a pre-clinical phase frequently is performed on animals. This cannot be considered part of a clinical trial *per se*. The results, although valuable in determining the safety profile of a device, may be extrapolated to the clinical situation only within limitations. |

A proper clinical trial is organised in a strictly methodical way and aims to answer one or more relevant and well-defined premises with respect to the clinical application of the device. The hypothesis has to be formulated prior to setting up the clinical study.

Clinical trials may be performed for different reasons. The intention may be for example, to test a purely scientific hypothesis, for instance, comparing the thromboembolic rate in relation to different anticoagulant regimens in patients with a heart valve prosthesis. When designed and performed accurately, the outcome will determine which regimen is the most beneficial to the patient.

A clinical study may also be done to suit regulatory objectives, primarily to obtain registration for a device in one or more countries. Examples of these are Food and Drug Administration (FDA) approval in the United States and the CE marking process in the European Union (EU). These studies focus in particular on proving that the clinical performance of a device and its patient safety profile are at least comparable, if not superior, to existing clinical approaches that are either therapeutic, prophylactic, or diagnostic in nature. Lastly, a study may be performed for reasons of medical economy or commercial purposes (see chapter 1).

CLINICAL EVALUATION IN THE DEVELOPMENT OF A DEVICE

In the various stages of its development, a device is taken through a number of clinical phases after pre-clinical testing has been successfully completed. In the initial stage of clinical testing, only a small group of patients will be exposed to the device. This type of trial is known as a *feasibility* study or a *pilot* study. The idea is to collect data that demonstrate a reasonable level of clinical performance and patient safety, and to allow clinical testing on a larger scale.

Although an acceptable safety profile may have been demonstrated in pre-clinical testing, this does not imply that the results may be unilaterally transferred to other test situations. It has to be realised that the feasibility phase is the first time that human subjects will be exposed to the device. The safety profile of the device in human application can be predicted reasonably well from an interpretation of the pre-clinical findings, but the safety profile is still to be refined. For that reason, monitoring is to be done intensely (see Table 2.1).

To demonstrate that the clinical characteristics of the device represent an appropriate level of performance and safety, the next phase of clinical testing is done on a larger group of patients. The number of subjects in this phase depends on either statistical considerations or regulatory requirements.

> **NOTE** For some medical devices, guidance documents have been defined for use by the parties involved in performing pre-clinical and clinical studies. These guidances are not mandatory, but it is strongly advised to follow them. Examples include FDA guidance documents and the ISO/CEN standards.

In contrast to drug studies, clinical testing of a device does not necessarily have to be done according to a comparative setup. For regulatory approval purposes, devices are commonly tested as individual entities not directly correlated to comparable technology. There is, however, a growing interest with regulators and the medical community toward randomised, controlled trials.

In Phase III, when approval has already been obtained for a device, the continuation of data collection is not uncommon, especially for high risk devices. At this point, the efficacy and

Table 2.1. Clinical Development of a Device

Clinical Research	Evaluation Synonyms	Study Properties
Phase I	Feasibility study Pilot study	Small group of patients (10–20)
Phase II	PMA study Clinical evaluation	Large group of patients (100–200 or more)
Phase III	Post-marketing surveillance Vigilance Registry study	Large patient groups Limited number of parameters

safety of the device have been demonstrated and are known to the clinician. In Phase III studies, emphasis is put on long-term follow-up of the subjects, with a particular interest in the complication rates (vigilance). In addition, studies may be set up to look into specific clinical characteristics of a device or to answer distinctive clinical issues related to the device's application.

DESIGNS FOR CLINICAL TRIALS

The outcome of a clinical trial to a large extent will be determined by the efforts that have been taken to set up the study in accordance with valid scientific criteria. The aim of the design will be to eliminate prejudice or bias that may exist with the investigator and others toward the device's clinical properties, and by doing so allow for objective testing as much as possible.

When the methodology applied in a well-designed clinical study is in line with sound scientific principles, the clinical research results obtained in only a limited number of patients may be extrapolated to the general population. This is precisely what is intended in a clinical trial: to test a hypothesis in such a way that the results will only in a negligible way (preferably not at all) be influenced by bias. This will lead to a situation where the outcome of the study will have a broader application than just for the patient group that participated in the research. Thus, it will be obvious that it is fairly difficult to overestimate the importance of proper design in clinical research. Various design formats are applied in studies with devices.

Over the years, a barely justified practice has grown in clinical research to reserve the term *study design* exclusively for a specific part of the overall design process, the intent to give a complete outline for a study. It can be noticed frequently that, unintentionally, reference is made to particular design formats, while actually it is meant to describe the overall design itself.

Terms like *prospective* or *comparative* studies do not express what overall design has been used in a study. Instead, they only represent certain specific qualities in the outline of the study. Although these are important features, they will only describe part of the entire study design, and, as such, they should only be considered *design characteristics.*

Design in a strict sense, however, refers to creating an extensive outline that covers the whole setup and execution of a study. For this reason, it has to be seen in a much broader perspective than simply implementing certain design characteristics. An example of what is covered by a study design in general is given in Table 2.2.

HOW IS THE PROCESS OF STUDY DESIGN ORGANISED AND CONTROLLED?

Usually, specifying and organising the study design is the next logical step to be taken after the definition of the central or core idea, which requires testing in a clinical setting (i.e., when the hypothesis has been formulated, the design may commence).

| NOTE | Defining the hypothesis is a very important activity before the clinical trial starts. However, it may be necessary to justify the trial for institution or company management approval. This may very well turn out to be *the* crucial issue, no matter how sensible the study may be. Budgetary constraints are known to be notorious obstacles! |

Various sources for initiating the definition of the hypothesis may be identified. The original concept for the trial may have emanated from clinical experience. The day-to-day clinical routine places the clinician in a favourable position to identify the optimal clinical application for a newly developed device or to indicate alternative applications for existing technologies. As such, the clinician's opinion should be valued highly.

The driving force behind the initial concept for the trial may also come from other sources, such as R&D staff or other employees in a clinical device company. Wherever the idea for the trial derives from, the design should always result from the input of various disciplines. Designing a study is a TEAM effort! The factual result of the design will be condensed in the study protocol.

Table 2.2. General Overview of a Study Design

Objectives of the Study	Defines what is actually intended with the study
Rationale for the Study	Forms the hypothesis that is being tested
Design Features	The study can have a comparative setup, be randomised, etc.
Patient Recruitment	Details the inclusion and exclusion criteria
Sample Size	Specifies the number of subjects required to test the hypothesis
Length of the Study	Duration of the study, per patient and overall
Number of Centres	Details the number of participating study sites based on sample size
Randomisation Procedure	Describes which method of patient allocation will take place
Statistical Procedure	Describes the method of data analysis
Parameter Selection	Defines the minimum amount of data that will be collected in each patient
Efficacy/Safety Assessment	Specifies the key parameters that will allow the performance and safety profile of the device
Monitoring	Describes the data collection and audit process to ensure that the study is run according to scientific principles
Case Report Forms	Intended to structure and control the complexity of data collection

NOTE | The design of the study will demand that the input comes from various sources (clinicians, R&D, regulatory, etc.). This will allow the defined study goals to cover what is actually intended with the device in the clinical setting.

Planning the design is a careful managerial process with two basic essentials: *proper organization* by the design leader and *good communication* in the team. Should these issues not be emphasised strongly, important details in the design may be easily overlooked. This may later lead to potentially detrimetal results when the study is being executed, when changes are difficult to implement. Once the design process has started, pay attention to the following issues to keep the pace going.

- Organise frequent meetings with everybody who is involved in the project to discuss progress and start up new initiatives.

- Make use of a list of essential items that have to be addressed in the meeting, to direct the discussions and streamline the debate.

- Make the minutes of the meetings always immediately following the meeting, when details are still remembered, and communicate these with all involved.

DESIGN CHARACTERISTICS

In clinical studies with devices, design characteristics are applied. These represent specific qualities in the outline of the study, but do not describe the entire design. Sometimes the term *statistical design* is used synonymously to design characteristics. The description of these features usually refers to clinical trials with drugs, but they also cover device studies. A brief overview, which does not claim to be complete, follows.

A study can be set up in a **prospective** way. This implies that the subjects enter a clinical study according to well-defined criteria and, following the clinical intervention that is being tested, will be followed up for a given length of time. During this period, data regarding the efficacy of the intervention and adverse effects in relation to it are collected and analysed.

| NOTE | It is essential in a prospective study that a strictly *pre-planned* approach is applied, which focusses on a standardised method of observation in the study. All subjects entering the study will follow the same procedures. The advantage will be that bias is significantly reduced, which augments the reliability of the study outcome. |

In contrast to this, a study can be considered **retrospective** when historical data are collected in patients who previously received treatment with a particular device, but in whom follow-up was not performed according to a planned, standardised method, as is customary in clinical research.

| NOTE | The fact that no planned observational strategy is applied in a retrospective study results in patient treatment variability, which may lead to serious potential bias. Hence, the outcome of a retrospective study is unreliable and should not be used. |

In a **comparative** trial, several well-defined patient groups, at least two, which have been subject to different treatments, are followed up according to a standardised approach. The intention here is to reduce or eliminate as much as possible the likelihood of differences existing between the various participating patient groups. By following this approach, the patient groups and the follow-up procedures applied become more or less similar; variability can be detected when the clinical results of the different groups are compared, which will be attributed to the differences in treatment.

| NOTE | Synonymous to the comparative clinical trial, the term **controlled clinical trial** is used. Patient groups treated differently are compared. One group receiving standard treatment functions as a control group for the other groups, who receive the experimental treatments. |
| | Previously, in controlled clinical trials with devices, sham procedures have been performed as a control to the actual experimental treatment. Today, this not considered ethical. Control usually refers to the standard therapy or procedure. |

In a **randomised trial**, the crucial idea is that subjects are not assigned to a particular treatment, but enter the study as they present themselves. The process of random assignment implies that subjects are enrolled in the study in an arbitrary fashion—the assignment is governed by a chance process in which the probability of being assigned to one treatment or to another is known or can be determined.

| NOTE | A randomised trial per definition will always be a controlled trial, as there need to be at least two treatments in which to assign subjects. |

Quite commonly, randomisation is performed in clinical trials with drugs. For device studies, this design characteristic has only recently been introduced. An important reason for this phenomenon is that in the majority of trials, only one group of patients, without a control group, was studied/treated with the device. More important, however, is that regulatory bodies, particularly the FDA, are becoming increasingly convinced that randomised studies will result in more valid results, less influenced by bias, when compared to non-randomised studies.

Today, also, in the clinical evaluation of devices, the **randomised, controlled** approach is considered to lead to the most reliable outcome. The setup implies that at least two treatment groups are compared, and the subjects are assigned to one of the treatments according to a randomisation scheme. Not until a subject has been determined to be eligible for the study will the group to which the subject has been allocated be revealed to both the investigator and the subject. Bias is eliminated in this way as much as possible.

In drug studies, **blinding** is often applied as an additional design feature in the randomised, controlled approach. Neither the subject nor the investigator will be informed about the treatment received by the subject until analysis of the collected data can be done or until particular circumstances make it necessary to reveal the treatment. This procedure results in an even lower level of bias.

Essential, however, is that the different drugs applied are indistinguishable from each other in order to avoid breaking of the blinding. It may be obvious that in trials with devices, blinding can rarely be used, as the treatment applied in a subject will usually be known to both parties. When blinding is not applied as a design characteristic (i.e., treatment will be known to the investigator and the subject), the clinical trial is called an **open clinical trial.**

In a **multi-centre clinical trial**, two or more clinical sites participate in the trial. All centres apply the same study protocol, and a data co-ordinating centre arranges the collection, pooling, and analysis of the clinical data. Patients in these studies are usually randomised to reduce bias. A trial in which only one site is involved is called a **single-centre trial**.

Both trial designs have particular advantages and disadvantages (see Table 2.3). Most important in the multi-centre trial is that it is reasonable to expect a larger number of patients to be enrolled, which will make it easier to satisfy the sample size set for the study. The downside, however, will be that the study population has been recruited from a number of centres and will probably be less uniform when compared to the single centre trial population. The lesser degree of uniformity will lead to an increase in bias in the multi-centre trial.

On the other hand, extrapolating the outcome of a trial performed in a single centre to the general population will be more difficult. There is always the possibility of centre-specific conditions that may bias the study.

Although difficult to apply in device studies, the **crossover** design is also worth mentioning. Subjects in the trial are initially assigned to one treatment and then switch to one of the other study treatments. Not only can the various study groups be compared, but also the individual patient becomes his or her own control. It should be obvious that in a trial in which two implants, for instance, artificial hips, are compared, a crossover will not be practical.

Sometimes, however, the crossover approach can be quite useful when comparing device treatments which are easily interchangeable between patients, and which will be applied to each patient more than once during the treatment. An example could be a trial comparing two different intravenous catheter designs.

A special situation arises when interim analysis during the study demonstrates a definite superiority of one treatment over the other. From an ethical point of view, crossover may then need to be applied.

Table 2.3. Single-Centre Trial Versus Multi-Centre Trial

	Single-Centre Trial	Multi-Centre Trial
Number of Subjects	Few	Many
Number of Centres	One	Two or more
Study Protocol	One	One
Data Centre	One	One
Organisation	Limited number of contacts in just one site facilitates the study setup.	Complicated level of logistics that increases with the number of centres
Investigator Meeting	Usually not done	Necessary to create group feeling and obtain consensus between the investigators
Communication	Relatively easy due to limited number of contacts	Controlling the information flow between the centres is a complex task for the monitor.
Recruitment	The limited patient pool may make it difficult to enroll a sufficient number of subjects.	More optimised enrollment conditions
Monitoring	Relatively simple	Complicated; the monitor manages the various centres so that the same protocol is followed in all centres.
Bias	There is a higher chance for a centre-related bias to form.	Due to the mixed study population, bias tends to be less.
Conditions	A small number of patients need to be enrolled in the study.	A large sample size is necessary in the study.

REFERENCES

Bourgeois, M. P. G., and P. Auclair. 1995. A Guideline for European Clinical Investigations of Medical Devices in Human Subjects. *European Journal of Cardiac Pacing Electrophysiology* 5:223–234.

Clinical Investigation of Medical Devices. 1994. Draft international standard ISO/DIS 14155.

Clinical Investigation of Medical Devices for Human Subjects. 1993. European standard EN 540.

Council Directive 90/385/EEC. 1990. Clinical Evaluation, Annex VII. *Official Journal of the European Communities* 189(17).

Council Directive 93/42/EEC. 1993. Clinical Evaluation, Annex X. *Official Journal of the European Communities* 169(1).

Meinert, C. L. 1986. *Clinical Trials: Design, Conduct, and Analysis.* Monographs in Epidemiology and Biostatistics, vol. 8. London: Oxford University Press.

Pieterse, H. 1994. Clinical Investigation with Medical Devices. *The Regulatory Affairs Journal (Devices)* May.

Pocock, S. J. 1983. *Clinical Trials: A Practical Approach.* Chichester, UK: John Wiley & Sons.

The Scrip Guide to Good Clinical Practice: Documenting Trials to U.S. FDA Requirements. Richmond, Surrey, UK: PJB Publications Ltd.

3

Managing Multi-Dimensional Clinical Projects

Herman Pieterse

Profess® Medical Consultancy BV
Heerhugowaard, The Netherlands

Clinical investigations with medical devices have become more and more complex. This implies that the co-ordinator of a clinical trial becomes more like a project manager with managerial responsibilities. Any clinical project can be defined as a group of functions performed by organisations to achieve a product (objective). The key element of this definition is the product. In our case, this can be translated to the objective: The project should be completed on time, within budget, and in compliance with appropriate technical standards or qualifications.

The basic principles of management are both complex and simple to explain. The key variable is PEOPLE. In order to learn the basic principles, one has to learn how people act and react. This involves human psychology, but sometimes the behaviour of animals also. Although a professional education and adequate training within the industry are essential, knowledge of the psychology of man in general and of the workers in the medical technology industry in particular is essential in order to be a successful manager of a clinical project.

The functional names for employees involved with clinical research range from study co-ordinators to research associates to study monitors. For readability reasons, the term *co-ordinator* is used for all of these names and functions.

MANAGERIAL SKILLS OF THE CO-ORDINATOR

The job of the co-ordinator of the clinical trial is to *get things done through people*. Two basic functions have to be fulfilled.

1. Cause decisions to be made.

2. Assure that actions are taken to implement the decisions.

To perform these functions on a project, the manager, independent of the level he/she is at in the organisation, is given varying degrees of *authority* and *responsibility* while being held *accountable* for the project. In order to perform these tasks successfully, he/she has to gain *respect.*

- *Authority* may be defined as the right (or power) to require others to perform or not perform activities deemed appropriate. This may either be given from above or granted from below.

- *Responsibility* is the obligation exacted to perform the agreed-on duties.

- *Accountability* is the liability for the proper discharge of the duties.

- *Respect* is the motor oil for success.

The co-ordinator will be appointed as the project manager with the objective to ensure that the project will be completed within the agreed time frame, within the budget, and to the technical qualifications. For this reason, the co-ordinator appoints a project team with all of the functions involved in a multi-dimensional case study. If the study is simple and straightforward, or if capacity is not available, then the co-ordinator will perform many of the activities themselves. Without any formal authority from above and trying to obtain informal authority from below, the co-ordinator is left alone to complete the project. This will tend to increase his/her personal risk. One should realise that this is a key variable that tends to make project management different from general (line) management. The co-ordinator has to get things done through his/her personal authority, drive, charisma, image, and the ability to obtain respect from others.

No matter the number of functions to be performed, the number of organisations (or individuals) involved, or the extent of the product to be delivered, the co-ordinator must provide the management of the project. This management can mean simple technical leadership and may be affected at any level of the organisation. As described above, the key variable in the management formula is people, and the management process is utilized to assist in *getting people committed and then motivated to reach the goal of completing the project on time, within budget, to the technical specifications.* This is really the challenge for a co-ordinator since it must be accomplished in an environment of change and risk. The management process includes planning, organization and staffing, and directing and controlling.

The co-ordinator should realise that he/she is expected to be fully informed, to take the lead, to inform the participants of the project adequately within time, to communicate optimally, to anticipate certain problems or pitfalls, to be the defender of the project to the top management of the company, to be the entrepreneur within the company, and to sell the project to management. This drive and these skills are prerequisites for a successful project leader. Organising the project systematically, in a step-by-step approach, and truly communicating can increase the probability of success for the project. Thus, the co-ordinator will recognise the risks and minimise the surprise.

ORGANISATION OF A MULTI-DIMENSIONAL CLINICAL RESEARCH PROJECT

A multi-dimensional project will consist of several subprojects. Numerous activities and subprojects have to be performed in order to achieve the objective, including

- preparing the investigational plan, including discussions with statisticians and experts in the field and management;

- preparing the Case Record Forms (CRFs) to ensure adequate data collection;

- selecting and qualifying investigators and staff;

- notifiying the authorities and Ethics Committees; and

- appointing a study team and arranging the budget.

A summary of the activities that have to be performed for a clinical study that complies with (GPC) for clinical investigations has been given in Figure 3.1.

Organisation consists of planning the tasks to be performed, the time necessary to achieve the goals, the capacity necessary, and the costs. The co-ordinator gives directions and controls the progress of the project. Organisation is, therefore, a continuous process of programming and implementing of the plan; of reporting progress against planning followed by control and adjustment of the original plan[1].

The co-ordinator makes the plan along with the other functions within the organisation. When the plan is complex, then a work breakdown structure has to be made. All activities must be described. The interrelationships between the activities has to be defined. Activities that should be performed before others can be started should be indicated. The critical path of the

---

1. It is obvious that managing a project is done with a strategic approach.

Figure 3.1. Profess® GCP Flowchart

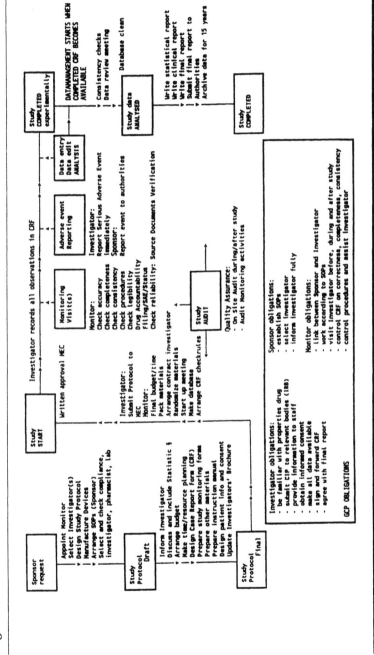

activities has to be identified. The alligators[2] or pitfalls have to be defined during the planning of the project.

Tools and Tips

Use a Checklist

In order to perform the tasks appropriately, a checklist is necessary to keep track of the project and to avoid forgetting certain activities. A sample checklist is included in Figure 3.2. In this clinical trial checklist, a compilation of the tasks is given.

The co-ordinator should follow a step-by-step approach, bearing in mind the following conditions: The co-ordinator and all other members of the team will work according to the respective standard operating procedures (SOPs). In order to guarantee quality and continuation of the project, the company has to always appoint an alternate co-ordinator for each project.

Design a Good Information System

Think of a logo or acronym for the project. Be sure that the members of the project team, their peers within the company, and management can identify the project. The information system should have a standard format. The type of system is not as important as its standardisation. Use a numbering system for proper reference.

Design a Standard Progress/Status Report

Every project should have regular progress or status reports. These can be weekly, monthly, or at whatever interval is best for the project.

---

2. A project is like crossing a river. The objective is to reach the other side of the river. In the river, there are several rocks. In project management terms, we call them 'milestones'. You can define the milestones to step on in order to reach the other side of the river. However, some of the rocks only seem to be rocks. But, in fact, these are alligators who will bite you and eat you. Try to take measures to check if the rocks are really rocks. This is where project management comes in: lobby, informally communicate, and know the information from the corridor. You never can define every rock. But every alligator that you discover before you step on it cannot hurt you. You are prepared and can take actions.

Figure 3.2. Sample Checklist for a Clinical Study

Subject	Item	Date Planned	Date ACTUAL	Check	Remarks
Investigator	Make selection criteria for investigators	|_|_|_|	|_|_|_|		
	Select investigators	|_|_|_|	|_|_|_|		
	Arrange for Pre-Study Visits	|_|_|_|	|_|_|_|		
	Check facilities and staff of investigator	|_|_|_|	|_|_|_|		
Documents	Collect updated and signed CVs from investigators	|_|_|_|	|_|_|_|		
	Collect normal laboratory values, if applicable	|_|_|_|	|_|_|_|		
	Arrange for an updated Investigator's Brochure	|_|_|_|	|_|_|_|		
Archive	Design an archive system	|_|_|_|	|_|_|_|		
	Design an investigator study file to be kept at the site	|_|_|_|	|_|_|_|		
Study manuals	Make investigator manual with information on logistics, procedures, and methods	|_|_|_|	|_|_|_|		

Figure 3.2 continued on next page.

Figure 3.2 continued from previous page.

Subject	Item	Date Planned	Date ACTUAL	Check	Remarks
	Make monitor manual on how to validate the data visually during monitoring	|_|_|_|	|_|_|_|		
	Make data handling manual on how to clean data in the database	|_|_|_|	|_|_|_|		
Analysis data	Design, test, validate, and approve database according to specifications	|_|_|_|	|_|_|_|		
	Agree on computer validation checks	|_|_|_|	|_|_|_|		
Notification study	Send notification to Competent Authority	|_|_|_|	|_|_|_|		
	Approval received within 60 days	|_|_|_|	|_|_|_|		
	Arrange import licence, if applicable	|_|_|_|	|_|_|_|		
	Import license arranged	|_|_|_|	|_|_|_|		
Study Devices	Order devices	|_|_|_|	|_|_|_|		
	Order any other study material	|_|_|_|	|_|_|_|		
	Pack and label devices according to protocol	|_|_|_|	|_|_|_|		

Figure 3.2 continued on next page.

Figure 3.2 continued from previous page.

Subject	Item	Date Planned	Date ACTUAL	Check	Remarks
	Make randomisation list, if applicable	|_|_|_|	|_|_|_|		
	Make decoding envelopes, if applicable	|_|_|_|	|_|_|_|		
	Send devices to investigator once all essential documents are in-house	|_|_|_|	|_|_|_|		
	Send other study materials to the investigators	|_|_|_|	|_|_|_|		
Instruction sites Initiation	Agree on an instruction/initiation visit with the site	|_|_|_|	|_|_|_|		
	Instruct investigator/staff	|_|_|_|	|_|_|_|		
Monitoring	Agree on frequency and tasks of monitoring	|_|_|_|	|_|_|_|		
	Agree on extent of source documents verification	|_|_|_|	|_|_|_|		
CRF tracking	Design tracking system for CRFs and other essential documents	|_|_|_|	|_|_|_|		

Figure 3.2 continued on next page.

Figure 3.2 continued from previous page.

Subject	Item	Date Planned	Date ACTUAL	Check	Remarks
	Design a stock control system for relevant materials (e.g., CRFs)	|_|_|	|_|_|		
Data handling	Decide on data entry procedure	|_|_|	|_|_|		
	Decide on data clarification procedures	|_|_|	|_|_|		
	Plan database clean	|_|_|	|_|_|		
Statistical analysis	Decide on which tests to be performed, why and how in Analysis plan	|_|_|	|_|_|		
	Perform analysis	|_|_|	|_|_|		
Reporting	Decide on format, layout, and contents of statistical report	|_|_|	|_|_|		
	Decide on format, layout, and contents of medical report	|_|_|	|_|_|		
	Decide on publication strategy	|_|_|	|_|_|		

Communication

Communicate news, problems from others, suggestions, and so on through the information system. One form of communication is to issue a newsletter from time to time. The newsletter can be more involved than the progress report.

Project Meetings

Organise meetings to report progress against planning. The agenda for the meeting should be discussed with the key members of the project prior to the meeting. The co-ordinator should try to identify all of the drawbacks, pitfalls, and problems that could be anticipated at the meeting. Perform the "lobby task" always.

To lobby is to invest time in learning the true motives of the people who could influence your project (hidden agendas). Explain your drives and goals. Try to find out what they are thinking about the project. Listen to what has not been said. Pay attention to non-verbal communication, the "atmosphere". Although this is the "political behaviour" of the co-ordinator, it is essential. The co-ordinator should be straightforward and honest to the person involved and to oneself.

Although it seems strange, through lobbying, the co-ordinator can manage to have the activity list even before the meeting starts. Discussions with individual members in the "corridor" can lead to an inventory of what they could commit to do. Then during the meeting, the co-ordinator can concentrate on the atmosphere, on brainstorming ideas, on involving quiet members of the project, and so on.

Always prepare the minutes of the meeting quickly! Within one working day, all team members should have a list of activities stating what is expected from them and by when. A simple activity list with subject, activities, responsible person, and deadline will suffice. An example is given in Figure 3.3.

Problem Solving

If problems arise, always follow the classical problem-solving approach.

- Identify the problem.
- Discuss with the persons involved the possible causes.
- Organise a brainstorming session (perhaps a discussion with only one person is adequate) on the possible solutions to the problem.
- Choose a solution.
- Execute the solution.
- Evaluate periodically whether the solution was the true solution for the problem.

Team Management

Manage your team members in a personal way. Try to have all people facing the same direction.

COMMUNICATION WITHIN A MULTI-DIMENSIONAL CLINICAL RESEARCH PROJECT

A co-ordinator is a junction within a communication network. There is always an objective for communication. The objective is to reach a mutual understanding and a business agreement.

Communication is a circle. The circle is complete when both parties listen to each other long enough so that they hear and understand the other sufficiently.

Discussion is often incomplete communication. Discussions and disputes can be limited or avoided by an adequate technique of conversation. To discuss a problem or item efficiently with a group, it is essential to first identify and define the problem. Suggestions for obtaining a better conversation are given below.

Ask Questions

Show attention to the problem. Pose the right questions at the right time. Continuously reflect on the purpose of the discussion: What are we doing? These are essential requirements for good conversation.

Pseudo-Perfection and Good Communication

Good communication can be achieved by giving the conversation the desired atmosphere with respect to feelings.

Pay Attention to Non-Verbal Communication

Most co-ordinators listen to the things that are said. This is important. However, the things that are not said are even more important. A person's non-verbal communication tells you what is meant. If you observe that there is a discrepancy between the things that a person says and what is communicated non-verbally (through attitude, the doubt in the voice, etc.), then you should check these words in an informal discussion with the person outside a formal meeting. It will help you to avoid stepping on an alligator that will bite you. We can learn how to do this from politicians.

Develop and/or Use Your Own Intuition

In a work situation, the co-ordinator has to perform the tasks and activities according to a systematic, structured approach. This is the basis for success. It will give the co-ordinator an overview of the time and the kind of activities that have to be performed. From time to time, he/she needs to get a helicopter view. Moreover, it will provide insight into the capacity necessary to perform the activities to the technical specifications and within the budget.

But next to these organisational aspects, communicative aspects clearly play an essential role. I refer to verbal communication in this respect—to information systems with the

Figure 3.3. Sample Activity List for Clinical Studies

Activity No.	Activity	Responsible Person	Deadline	Predecessor	Status	Comments
			_ _ _			
			_ _ _			
			_ _ _			
			_ _ _			
			_ _ _			
			_ _ _			
			_ _ _			

Figure 3.3 continued on next page.

Figure 3.3 continued from previous page.

CATEGORIES
Preparation Study

- Investigator related
- Organisation related
- Financial aspects

Execution of the Study

- Monitoring related
- Data management
- QA audit

Analysis of the Study

Explanation of the headings and instructions:

- A number is given for each activity.
- All activities are mentioned.
- The person who is primarily responsible to complete the activity is mentioned; other persons involved in this activity can also be mentioned.
- The anticipated completion data of the activity is mentioned.
- The number of the activity that must be completed before this activity can start (to inform others on the interrelationship of activities) is given.
- All remarks can be outlined that will elucidate the activity; once the activity is completed, the word COMPLETED is written.
- The next status report will mention the completed events in another section named COMPLETED EVENTS.

objectives to inform the participants of a group, team, or project optimally with respect to the progress of the project. I also refer to non-verbal communication.

Intuition is

- a seventh sense that you can develop to help you read between the lines, to listen to what has not been said, but is surely meant by someone;

- learning to assess a situation, to feel the atmosphere, attitudes, behaviour of people; and

- recognizing that psychological processes are the basis of the reactions of people—how they will react in specific situations—and paying attention to the atmosphere. These psychological processes and the reactions of people are related to the company culture.

Develop and use your intuition to assess certain situations. Use it to determine the strategic options that emerge from the analysis of a problem or conflict; also use it in situations where people become euphoric with respect to the completion of a project.

Learn to listen to what has not been said. Work from your own basic philosophy: Remain honest and sincere!

Sometimes this means that the truth must be said even when it is painful. But remember, the truth may hurt, but revealing the truth at a later time is much more painful.

Listening

As stated before, discussion is often incomplete communication. Communication is achieved by listening carefully. Communication is continued by listening. Communication ends when listening stops. *The co-ordinator should listen especially to the things that are not said, but meant.* There are many suggestions for good listening and communication.

Attentive Listening

Improve a conversation by attentive listening. Show interest and understanding for the subject someone wishes to discuss.

Be creative and try to find solutions. Determine your point of view with respect to the subject. Show interest and communicate verbally.

Questioning

Use the proper kinds of questions. There are suggestive questions (but be careful with these because they are quite manipulative), direct questions (for non-critical issues), reflecting questions that summarize the issue, open questions (provide the opportunity to obtain more background information about the subject), and multiple choice questions that give someone the opportunity to choose between options (the backdoor principle).

Reflect Continuously

Why am I communicating? About what do I communicate? How, with whom, where, and when do I communicate the best?

Program Meetings and Discussions

Prepare the meeting. Try to identify the "unidentified alligators". Try to identify problems that live below the water line.

A meeting is often a forum where people express their status in the project and in the organisation. Give people space, but anticipate. "Take care of the show" element of a meeting. Above all, stick to the agenda. Be a good leader: result oriented, charismatic, motivated, passionate, and diplomatic. Don't be too pushy. Neutralise negative influences. Summarise between the discussion points and repeat the decisions.

Meeting Minutes

After the meeting, prepare the minutes in the form of an action list with annexes of the presentations, if applicable, within the shortest time possible. Strength lies in speed. The minutes of a meeting are a powerful instrument for the project leader. However, use this power to be honest to both the team members and to yourself.

The Critical Path

Always communicate the "critical path" to the members of the project. Convince them that it is necessary that they perform the activities that are on the critical path. Share with them the "overview" you have over the whole project (i.e., the helicopter view). Above all, agree with them that they should inform you in case they cannot keep their promise to perform the critical activities on time. *It is better to be informed that something cannot be done on time, than not to be informed at all.*

Problem Solving

Never avoid or negate problems from others. Treat complaints objectively. Be creative in finding solutions for problems.

EFFECTIVE MANAGEMENT

Effective project management, but also process management, can be performed if two main requirements are met: good organisation and good communication. Both can be achieved if proper choices are made for the political–tactical, strategic, and intrinsic aspects of the project. By means of a good lobby, the co-ordinator can obtain a proper "commitment" from the team members. Moreover, it will also positively influence the management of the company.

Full consensus on the objective of the project is essential to the success of the project. However, each member of the team should be fully aware of the general phases of project management.

Phase I: Enthusiasm Phase

Each project always starts with this phase. The co-ordinator has no problems. He/she is recognised as being successful.

Phase II: Dissolution Phase

A problem arises that jeopardises the planning. Other competitive projects which are in their "phase I" stage have been initiated. The company looks at the co-ordinator for solutions. Here, the right skills must be present, and actions have to be taken: problem solving, lobbying, structuring the problems, finding solutions, mobilizing the team members, and so on.

Phase III: Panic Phase

If, for one reason or another, the co-ordinator cannot cope with the problems from phase II, then panic starts. People will discuss that the project is a lost case, that the product cannot be registered, that they always said so, and so on. In other words: EMOTIONS get hold of the project. Moreover, other projects may have higher priority at that moment.

It is the task of the co-ordinator to listen to signals of panic in the "corridors" and to take appropriate actions as soon as possible. Because the project works with people, the members of the project become tired and disappointed. Being conscious of these symptoms, the project leader learns to identify them. This phase is not an irreversible phase. If acted on accordingly, the co-ordinator now has the opportunity to show that he/she is a true leader. Be creative. Clear the air thoroughly by discussing the problem out in the open. If, for any reason, the co-ordinator does not succeed, then someone becomes the "scapegoat".

Phase IV: Initiation of Blame

The company indicates a person to get the blame for the negative results. Now it becomes tricky. The organisation wishes to appoint a scapegoat. It is not always the project leader. This depends on the credit he/she has built within the organisation. The co-ordinator who is responsible for the progress of the project should immediately neutralise. There is a fire! This is the last chance.

This phase occurs at least once in the life cycle of a project. There is no more fun in leading the project. Always defend the team members to third parties (like a parent would do for his/her children). With the respective team member, you can discuss the problem, which sometimes means that the absolute truth must be said. In most cases, this phase takes place in informal spheres—in the corridors, in the gossip circuit. It is my experience that women are much more honest to you and themselves than are men.

Take your advantage but do not interfere in this "guerrilla circuit". Leave it as it is. In phase IV, there will be no more involvement of any of the team members. A large proportion of the development projects within the healthcare industry, but also projects to register a product already developed, end in this phase. This is not typical for small companies. Large companies are notorious with ending projects in this phase. Most of the time, you will hear that the reason for ending the project is "the product failed us". In many cases, the true reason is a lack of proper communication and management skills.

Phase V: Analysis

Analysing the cause of the problems and punishing the innocent often occur in this phase. Third parties will invest a lot of time in analysing the problem and its causes, and pushing forward persons whose competence has not yet been established. Many projects are put on hold in this phase. The project leader is isolated and left alone. Innocent secretaries or others will be punished.

If due to purely external circumstances and factors the project is resuscitated and lives further through its phases without any jeopardy, then phase VI arrives.

Phase VI: Rewards

The person who has not been involved in the project is rewarded. The product is on the market. Many people in the company will claim the credit. Due to their contribution, the project has now become a success.

The co-ordinator, but also the team members, should always be aware of these phases of the project. Together you should realise in which phase a project is. Try to look at the project with a helicopter view. When every team member is conscious of the phase of the project, phases III to V can be avoided.

PROJECT TEAM MEMBERS

The clinical co-ordinator has appointed the members of his/her team in a multi-dimensional clinical project. The team will now perform its activities within the matrix of the organisation. All members will usually be appointed only part-time to perform their tasks for the project. If the co-ordinator organizes a meeting, he/she can ensure that each member can give an equal contribution to the meeting. During the course of the project, the co-ordinator will meet several kinds of team members:

- the person who knows it all
- the quiet listener
- the expert
- the troublemaker
- the politician
- the pessimist
- the optimist
- the dreamer
- the person who can do it all

The co-ordinator has to deal with all of these characters. Sometimes the co-ordinator will be the schoolteacher; sometimes democratic, often decisive. He/she must develop leadership style into a situational one.

PROJECT MEETINGS

The co-ordinator will make sure that there is a proper climate for the meeting; he/she should neutralise any disturbances and control the circumstances. The meeting should be properly prepared, and be neutral with respect to the outcome of the discussions but not to the process. The objective of the meeting will be met. At the beginning of the meeting, the co-ordinator assures that each participant agrees on the objective of the meeting, gives room to air emotions, and assures that the process is in line with the objective.

During the meeting, the project leader

- periodically summarises the meeting up to the point where the meeting is and

- stimulates participation of all team members (e.g., by putting questions to the listeners, supporting, informing, involving, and slowing down).

In conclusion, the project leader is both a police officer and a referee, whichever is suitable. The project leader is honest and neutral.

TIME PLANNING AND SCHEDULING
IN CLINICAL RESEARCH

Once the co-ordinator has appointed the team members, he/she will make a proposal for the project plan, a true scenario.

- Describe the objectives, deadlines, and completion dates.

- Detail the functions to be performed, thus defining "what" is to be done.

- Decide which organisations or individuals will be responsible, thus defining "who" is to do "what".

- Plan the way in which the work will be accomplished, thus defining "how" it will be done.

- Establish a time plan and schedule of the work, thus defining "when" the work will be done.

- Establish a resource plan and budget for the project and its pieces, thus defining "how much capacity" it will cost.

- Communicate the plan of who is to do what, how, by when and for how much to all the parties involved and to management.

A complete list of all activities—a scenario—has to be made, with an indication of the status of certain activities and who did the work. The list should be as complete as possible, so that the interrelationship of the activities and the predecessor of all activities are defined. This defines the "critical path". Some of the activities can be performed in parallel; others have to be performed in serial order.

SUMMARY

The co-ordinator will have the following functions.

- Provide leadership and direction.
- Interpret project requirements.
- Define and assign responsibility.
- Guide the project planning.
- Monitor the progress of work.
- Act as a liaison with management and/or the sponsor and the workers.

To accomplish all of these functions, the ideal co-ordinator should possess the following skills.

- managerial
- technical
- project planning and control

- financial

- marketing

- negotiating

- communicating

Since it is seldom possible to find an individual who embraces all of these skills in the right amounts, the co-ordinator must have one other key skill: knowing what he/she does not know and identifying what resources can be utilized for assistance. The co-ordinator is only one member of the project management team. All team members and functional management should complement the co-ordinator.

Some of the key detailed functions the co-ordinator must perform are as follows.

- Monitor and review the compliance of technical work to the specifications at design and final product.

- Review and decide the staging of change and the wisdom of desired (as opposed to required) change.

- Establish planning guidelines and monitor planning.

- "Sell" the project to management, the sponsor, and the organisation.

- Administratively support the work teams.

To become an efficient study co-ordinator, it is necessary to learn to deal with a number of instruments, to learn a few tips. Many of these tips were given in this chapter. The basic principles of management were discussed. With examples, some insight into the practice of clinical research and in the behaviour of fellow workers in the pharmaceutical industry has been given. Be aware of the principles, and remember the following.

- Use your intuition.

- Listen to what has not been said.

- Be a communicative listener and show interest in others.

- Be informative.

- Structure your activities.

- Be systematic and strategic in your approach of reaching your objectives.

Section II

Strategy and Project Planning

4

Device Classification and Risk Assessment

Marja G. de Jong

Medidas
Wijk bij Duurstede, The Netherlands

Peter Duijst

Aortech Europe Ltd.
Almere, The Netherlands

INTRODUCTION

Regulatory systems in general define classification systems for medical devices, in such a way that the most stringent regulations apply to those medical devices which in clinical use entail the most potential hazard for patient and environment. The philosophy behind this principle is fairly simple and logical. Taking low-risk medical devices through the same rigorous testing that is obligatory for high-risk devices would likely create unnecessary, lengthy regulatory approval procedures and consequently generate a significant level of cost-ineffectiveness in socioeconomic terms, without effectively enhancing patient safety. Ultimately, the overriding objective for the classification of medical devices is to match patient safety assessment with a graduated system of regulatory requirements.

An integral part of the safety assessment of a medical device is the identification of possible physical injury and/or damage to health or damage to property that could result from the clinical use of that device. The regulatory standards define this qualitative process as *identification of hazards*. Once the process of recognising hazards has been completed, quantification of the risk of each separate hazard occurring may start. The probable rate of each hazard's occurrence will be estimated, and expressed as a percentage.

From the aforementioned it is clear that the classification of a product can be done only once proper risk assessment has been performed. In most regulatory systems the hazard identification and risk determination of the product determines its classification category. Although the regulatory world is currently moving towards a harmonised system, it is important, nonetheless, to realise that the existing varying regulatory requirements may classify the same device differently. This could have serious implications, for instance, in whether or not there is a need for a clinical investigation.

This chapter discusses classification and risk assessment with specific emphasis on their relationship to clinical investigation.

RISKS AND HAZARDS

Identification of hazards—both those that are inherent in the device itself and those that are associated with the device when used—lies at the heart of risk analysis. Hazards are or user or property. A sample of possible hazards are found in Table 4.1.

To identify the hazards pertaining to a product requires basically nothing more than making a list of all the technical and clinical characteristics of the device, accompanied by those events that could occur with the product when used in a clinical setting.

Table 4.1. Sample of Technical/Clinical Hazards

Energy Hazards

- Electricity
- Heat
- Mechanical force
- Radiation
- Moving parts

Biological Hazards

- Biocontamination
- Bio-incompatibility
- Toxicity

Environmental Hazards

- Electromagnetic interference
- Incompatibility with other devices
- Contamination due to waste products

User-Related Hazards

- Inadequate labelling
- Inadequate patient information
- Inadequate specifications
- Reasonably foreseen misuse/abuse
- Incorrect measurements
- Erroneous data transfer

Functional Hazards

- Inadequate performance
- Loss of integrity
- Improper re-use/abuse
- Deterioration in function

Information concerning the identification of hazards, as well as the quantification of the hazards, can be obtained from various sources including:

- published standards;

- scientific data;

- clinical data from literature and trials; and

- post-market surveillance data including complaints and adverse events.

| TIP | Some of the hazards identified will be clinical hazards, which will need to be addressed through clinical assessment. |

Once this qualification has been done, the severity of each hazard's occurrence has to be determined, in order to understand whether the risk is *catastrophic, critical,* or *negligible*. In other words, each hazard is to be 'quantified'. This quantification of the degree of severity to health and safety is known as **risk estimation**. Similarly, the acceptability level of each risk's occurrence when weighed against the benefits of the device has to be determined.

The likelihood of the occurrence of the hazard could be:

- frequent

- probable

- occasional

- remote

- improbable, or

- incredible

The degree of severity of the risk is based on a 'worse case scenario' (i.e., determining the worse possible mishap that could occur when the device is used).

| TIP | FTA techniques are *deductive* methods working 'top-down' and are used in design control; FMEA techniques, or *inductive* techniques or 'bottom-up' techniques, are applied when the design has become more detailed. |

| NOTE | The manufacturers of medical devices are legally required to do a risk analysis. The harmonised standard (EN1441) does not relieve the manufacturer of its obligations to conduct a risk analysis. Furthermore, the standard does not mention levels of acceptability; this is left to the manufacturer. The manufacturer must identify these acceptability levels, which should not be in conflict with the opinions set by the Competent Authorities (CA) or Notified Bodies (NBs). |

Risk analysis techniques are applied for quantitative or qualitative estimation of probability in relation to the possible consequences that could occur when the device is used. These techniques may include Fault Tree Analysis (FTA) and Failure Mode and Effect Analysis (FMEA)[1]. Most importantly, the tool used to quantify the hazard must be appropriate to the hazard.

Risk Management

The entire process of overseeing and controlling risks involved in the use of medical devices is known as **risk management**. Figure 4.1 gives an overview of the various elements of importance which this process comprises.

The management of the risks is critical to the assurance of product safety during the lifetime of the product. It is an integral part of the design process. Essential in risk management is that the product needs to be free of any *unacceptable probabilities*.

Figure 4.1. Risk Management

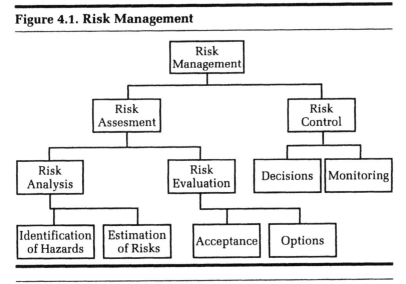

1. FTA is a deductive technique that uses the potential consequence as its basis. Lower level fault conditions that could cause the consequence to come into being are identified. FMEA is an inductive technique by which the frequency and consequences of individual fault modes as they relate to product functions are systematically identified and evaluated.

Risk Analysis Overview

Risk analysis distinguishes between aspects related to:

- the intended application and use of the device,

- technical solutions, and

- characteristics specific to the device (whether inherent or associated).

Certain risks, however, can basically be considered inherent to a device (e.g., capsular contracture in mammary protheses). Until techniques are developed to control such risks, they have to be accepted provided they do not compromise the safety of the patient.

Clinical data and existing knowledge of clinical performance of a comparable device may be an important component in risk analysis. Historical clinical data can justify the acceptance of risks or may lead, by way of clinical assessment, to a reduction of risks associated with the device. The objective is always to ensure that the risks are as low as are reasonably practicable. This is a premise critical in risk analysis. Table 4.2 gives an overview of the process of risk analysis.

Risk analysis is multi-disciplinary, covering many areas of expertise (e.g., engineering statistics, research and development, clinical affairs, manufacturing, marketing/sales). Like clinical investigation or managing a quality system, it needs to be done in a structured and disciplined manner. Undertaking risk analysis is part of the quality system process. Moreover, it is a process that is ongoing for the life-time of the product.

| TIP | Ensure that the risk analysis is updated periodically and reviewed. |

CLASSIFICATION

Once risks are identified and the process of risk analysis completed, the classification of the device can be determined.

Table 4.2. Steps for Risk Analysis

Characteristic	Tips and Hints
• Write a procedure defining the process of risk analysis.	• Include how hazards are to be identified. • Include methodologies by which risks are to be determined. • Include criteria to be used to determine acceptability of risk and to aid in choosing risk-reduction measures. • Define any risk-reduction methods to be used. • Define the decision-making process. • Define who is responsible for the risk analysis. • Determine how risks are going to be managed. • Define the risk analysis team. • Define how the risk analysis is to be documented.
• Elect a project team and a project leader.	• It is a multi-disciplinary task. Ensure that all parties are involved in the process from the beginning.
• Define the intended purpose of the device.	• Complete description of the device should be included in the risk analysis documentation, along with the intended purpose of the product.
• Identify all potential characteristics that relate to the device.	• List all of them no matter how mundane they may be. • Take into consideration any particular international requirements for the product or safety of the product. • Examine issues of intended purpose of the device; for whom the device is intended to be used; materials and components including biologicals; sterility; temperature and humidity factors; gases; interactions with other products; operational, transport, storage factors; restrictions on shelf-life; life-time of the product (ageing); etc.

Table 4.2 continued on next page.

Table 4.2 continued from previous page.

• Define any hazards which could be associated with the device or which are inherent to the device.	• Clearly state known technical and clinical hazards. • Identify unknown technical and clinical hazards.
• Define the scope of the risk analysis.	• Define the methods to be used for the risk analysis (e.g., inductive, deductive, other). • Identify any problems associated with the risk analysis. • Formulate objective of the risk analysis based on the main hazards identified. • Define what is being analysed, including any limitations. • Detail technical, clinical, and related conditions relevant to the product. • Make assumptions/hypotheses. • Identify any decisions that are to be made
• Estimate the risk associated with each of the identified hazards.	• Investigate the hazards identified using the appropriate methodologies. • Do a pathway analysis tracing the hazard from the source to the receptors. • Define the data requirements and how these are to be documented. Use of databases where correlation can be done statistically may be appropriate. • Examine standards, scientific data, post-market surveillance data, clinical data and so forth to help in risk assessment.
• Review each of the risks and determine whether they are acceptable or not acceptable.	• Determine corrective actions. Eliminate or reduce the risk of the hazard occurring. According to the requirements in most regulatory systems, risks are first and foremost to be eliminated. Should they not be able to be eliminated, then their severity is to be reduced to clinically acceptable levels

Table 4.2 continued on next page.

Table 4.2 continued from previous page.

	(i.e., to the lowest reasonable level practicable). Should they not be able to be eliminated or reduced, then appropriate warnings should be included in the labelling.
• Verify the findings.	• Formally review the risk process and document the review process. Testing and changes to product design may be necessary.
• Write the risk analysis report.	• Document the findings and the entire process.
	• The report depends on the objectives and the scope of the analysis. Include, for example: • title; • cover page; • amendment or history sheet with revision levels; • executive summary; • table of contents; • objectives and scope; • limitations/assumptions; • description of the product and intended use; • analysis methodologies used; • hazard identification criteria and the hazards identified; • description of the models used for the risk estimation and results • corrective actions; and • discussion of the results.

Logically, the more severe the risks a device entailed with its clinical application, the higher the classification assigned to the device and/or the more the need for clinical assessment. Classification in turn influences whether or not there is a necessity for clinical investigation. See chapter 1 for information on when clinical investigations are required.

It has all ready been established in the aforementioned section that classification of a device is directly related to the risks associated with it. The risks are related to various parameters (e.g., duration and degree of invasiveness) which form part of the classification system. Most regulatory systems maintain three classes of devices: Class I, for low-risk devices; Class II, for medium-risk devices; and Class III, for high-risk devices.

In some regulatory systems, particularly in the United States, for example, classification can be a long and tedious process. Many times, devices will be designated one class and later be re-classified into a higher or lower class. This reclassification is dependent on whether the product has been shown to be substantially equivalent to an existing product already on the market, and by acceptability of the device and its technology by the medical community and regulators alike.

Under the Code of Federal Regulations sections 800 to 812, the United States lists classifications of the medical devices and in-vitro devices (IVDs) that the Food and Drug Administration (FDA) has categorised. The primary criteria for classification rests with the anatomical area where the product is used. Under the amendments to the FDA requirements, most products, for which substantial equivalence cannot be demonstrated are nowadays put into Class III, where they are subject to clinical trial and post-marketing application. Should a company later prove substantial equivalence with existing devices, the FDA may reconsider the product classification. Table 4.3 gives an overview of the classification types.

Whereas the U.S. requirements focus on the anatomical area of application, the European legislators set up a system of classification based on a different set of criteria, including:

- the device's duration of contact with the body;

- the device's degree of invasiveness with the body; and

- the device's effects on the body—that is, local versus systemic.

Duration involves the period of time a device is in continuous use. For instance, a device can be used

- transiently (i.e., less than 60 minutes' continuous use);

- short-term (i.e., less than 30 days' continuous use);

- long-term (i.e., more than 30 days' continuous use).

Duration is linked to *invasiveness*—the degree to which a device penetrates the body, either through a natural opening or through an artificial opening created during surgery. Low-risk, non-invasive devices will most likely be categorised Class I.

Sensitivity of the part of the body where the device is used or applied is linked to *intended purpose of the device*. The more invasive the device and the more contact it has with vital organs, the higher the risk classification of the device.

NOTE The rule of thumb in classifying products in the European Union is that the more invasive a device is and the longer it remains in contact with the body, the higher the risk and the higher the classification.

TIP It is important to understand the classification rules in each regulatory system under which a product is to be marketed. Different regulatory systems may classify the same device differently, which influences whether a clinical investigation is to be conducted or not.

Comparative Classification Requirements

Differences between the classification criteria in the U.S. and European requirements meets do exist, as shown in Table 4.3. Nevertheless, two elements should always be taken into consideration from the outset:

1. Products may be classified one way in the European Union and another way in the United States. When a product requires clinical investigation or submission of

Table 4.3. Summary of Comparative Classification Rules (European Union and United States)

Class	European Union	United States
I	• Non-invasive devices that, for example, do not touch patient, or contact only intact skin, or are used for channelling or storing fluids for eventual administration, or are reusable surgical instruments	• Devices that either do not touch patient or contact only intact skin • Non-invasive devices used for channelling or storing for eventual administration to the patient
II		• Any device not in Class I or III (e.g., contact lens care products)
II-a	• A variety of devices. For example, may range from non-invasive to surgically invasive devices of a transient nature to monitoring vital signs.	
II-b	• Variety of devices ranging from non-invasive devices that modify biological or chemical composition of blood, body liquids, or other liquids intended for infusion, to longer-term implantable products.	
III	• High-risk devices which are life-threatening, have a potential for causing high risk of injury, or are used in the Central Nervous System (CNS) or central cardiac system (CCS)	• High-risk devices that are (a) life-sustaining; or (b) of substantial importance in preventing impairment of health; or (c) a potential for causing high risk of injury or illness, or (d) subject to a new ruling issued by the Food and Drug Administration (see Table 4.4).

an investigational device exemption (IDE) in the United States, it may not require such a clinical investigation process in the European Union, and vice versa.

TIP

This is a critical point when planning a regulatory/clinical strategy for the global launch of a product. (See chapter 5.)

2. The way a product is classified influences the level of clinical data required. For instance, for a Class III product, clinical investigation most likely will be necessary, as well as for any implantable product and long-term invasive product in Class II-a and/or Class II-b. The following table outlines the classes of product in a comparative manner. (4.3)

Classification and Conformity Assessment

Risk determination mandates the classification of the product, which in turn determines the conformity assessment or approval process a device will be subject to. The most rigorous requirements are applied to those products classified as having the highest risk. An overview of the classification categories as related to conformity assessment or approval is given in chapter 1 Table 1.2 *(Medical Device Classification table)*.

CONCLUSION

Safety of a device is defined as the freedom from unacceptable probabilities which could occur when the device is in clinical use and the consequences which might lead to endangering of the life or health of the patient or the user of the device. In order to justify the clinical acceptability of risks entailed by a device, the probability of the risks occurring and the potential clinical impact those risks may have need to be understood. This necessity makes a proper risk analysis imperative. Two sequential and critical steps are crucial when performing risk analysis. First, the qualitative process of identifying potential hazards associated with the clinical use of the device needs to be performed. Second, for each hazard identified, the probability rate of its occurrence needs to be estimated and expressed in a percentage.

Because the classification of any device is explicitly linked to the risks associated with its clinical operation, it is obvious that risk analysis plays a crucial role in the various classification parameters that are offered in the various regulatory systems. After a device has been classified, the clinical implications can be overseen. In other words, once it is understood into what risk class a device falls, the regulatory needs for clinical investigation can be determined. Again, these needs differ from one regulatory system to another.

In the overall process of making innovative design changes beneficial to the patient and successful after their introduction in the clinic, an in-depth understanding of the potential severity of the risks associated with their use is imperative. For that reason having a structured and controlled approach toward risk management is a strong tool in attaining this goal.

5

Clinical Strategy Development

Jaap L. Laufer
EMDC
Eerbeek, The Netherlands

More often than not, a variety of sometimes conflicting objectives require reconciliation before a study protocol can be approved. Through the careful preparation and planning of a clinical trial, strategy and plan conflicts can be identified, understood, and generally reconciled early in the clinical trial process. Many factors influence a clinical strategy, including

- type of device being investigated;

- classification of the device;

- regulatory requirements;

- risks associated with the device; and

- intended clinical use of the device.

There are also key questions that should be asked when developing a clinical strategy, including but not limited to the following:

- What are the reasons for the trial?

- What is the strategy?

- What are the objectives?

- What is the mechanism by which the trial is to be conducted?

Once these questions have been addressed and the objectives have been agreed to, chances are the company will achieve a successful trial. However, variables exist and influence the answers. For example:

- The size of the company is a major contributing factor. Is the company a small venture capital organisation or a large multi-faceted and complex organisation?

- The status of the device being investigated is also a major factor. Is the company investigating the design of new products, or modifications of existing products?

This chapter focuses on the strategic considerations that should be taken into account and that form a critical part of the clinical strategy and its subsequent written plan.

STRATEGY AND TRIAL OBJECTIVES

Strategy Steps

A regulatory and commercial review should take place on all design ideas at the design input phase. This is as applicable to new products under development as it is to modifications to existing products. The following considerations should be carefully assessed and justified.

- What is the intended use of the device?

- What is the medical justification (i.e., benefit) of the device?

- Which market segment will it be most beneficial to?

- Which medical specialists will use it?

- In what markets do we want to be? Why do we want to be there? When do we want to market the product?

- Are the goals achievable? (Usually a best/worse case scenario is a useful tool.)

- Do the current financial resources allow these goals to be attained? If not, are there other resources that will cover even the worst-case scenario?

- What are the detailed sequential steps that must be taken in order to achieve the goals?

All of these elements not only form part of the product file but also impact decisions pertaining to clinical trials. In cases when clinical studies are necessary, whether for commercial or for regulatory purposes, the clinical protocol as well as the overall study is subject to internal approval; authorisation by the appropriate persons (i.e., the medical director or clinical director) is a *must*.

Development of the clinical strategy is required for new products because it

- focuses the management on a conceptual pathway;

- helps avoid obstacles and complications, such as new injections of capital; and

- allows protocol amendments or design changes to be taken into account in advance.

For design changes to existing products, the relevant questions stated previously must be asked: Should clinical data be required? The answer to this query should be derived from the risk analysis, with verification by the independent advisor. The conclusion reached as to whether a clinical trial is or is not necessary—and for which reasons—needs to be adequately documented to ensure compliance with the requirements and the overall company objectives.

Independent Reviewers

The clinical strategy plan forms part of the technical documentation. It is therefore necessary to ensure that a consensus be reached on the document, with input from various departments, including Regulatory Affairs, Clinical Affairs, and Quality Assurance. Having the entire design process and the development of the clinical strategy subject to independent review is also recommended.

TIP	Have the draft clinical strategy plan reviewed by independent experts in Regulatory and Clinical Affairs as well.

The use of knowledgeable, impartial, clinically experienced individuals is not uncommon in the clinical investigation process. For clinical investigations, a medical expert should be part of the clinical investigation team (see chapters 3 and 6), and establishing a knowledgeable, well-focused sounding board of impartial, independent medical advisors should be considered. They should be involved from the beginning to the end of the process, starting with the design phase and prototype assessment. They may act as a permanent peer group. The objective of the independent medical advisor(s) is to avoid the making of incorrect assumptions and the drawing of false conclusions.

For products with long-term clinical data follow-up, the advisors should be asked to help draft a post-market surveillance protocol.

Costs and Company Sizes

Experience teaches us that in the vast majority of cases, the costs associated with new products, changes to existing products, and pending clinical evidence are grossly underestimated. The difficulty lies in there being no measuring-sticks—not even to help in the estimating of the costs. Therefore, the more research done, the better the preparation, and thus the fewer the obstacles. A rule of thumb is to calculate the bottom line, then increase it by 30 to 40 percent if investigating either a new device or a new use of a device, and by 20 percent if investigating a major design change requiring new clinical evidence.

TIP	*Do not* underestimate the design costs and the cost of the clinical strategy plan.

Small companies are quite often one- or two-product start-up companies, whether venture capital or otherwise. Capital expenditures may therefore constitute one of their biggest investments, if not be the major item. In such situations, failure to recognise all of the issues in terms of clinical trials can lead to delays to market, increased capital needs, and missed commercial opportunities. Understanding the strategy and the needs in these cases also helps smooth the relationship with any potential investigator.

Large companies, on the other hand, have a multi-layered structure with a complex decision-making process. In the majority of cases (but not always) they are well-capitalised and have a cash-flow adequate to support various in-house departments, such as Research and Development (R&D), and Medical/Clinical Affairs, as well as medical advisors and consultants.

Small companies tend to recognise the impact of design changes very rapidly. This stems from the fact that they are accustomed to a relatively rapid development of their product.

In a large company, such awareness occurs gradually. Even a gradually evolving design change, however, may not always be recognised as necessitating significant changes to the investigational plan, which could lead to false conclusions.

Regardless of the size of the company or how the funding of the company is arranged, taking into consideration all the elements of risks, issues, and related factors is critical to the product and the business. Such assessments can contribute greatly to recognising the need for additional clinical evidence. For instance, a new application of a product may not only command a design adaptation, but could also allow the product to cater to an entirely different group of specialists. All these elements need to be examined. A clinical strategy, like a business strategy, needs to remain fluid and evolve over time.

Capital needs are also part of the overall strategy. Any type of study to be conducted, regardless of the type of device or its intended use, includes but is not limited to the following types of capital costs:

- the cost of the protocol development; advisors; clinical report forms; and draft consent form;

- investigator selection (travel);

- legal/regulatory review;

- the cost of supplies;

- the cost of medical procedures, laboratory analyses, hospital stays, etc.;

- special packaging;

- special labelling;

- the cost of training and investigator meetings;

- on-site staffing (e.g., monitors, data co-ordinators);

- Good Clinical Practice (GCP) audits;

- data handling and costs;

- follow-up;

- reporting; and

- investigator fees.

Communication Obstacles

Apart from the capital needs, potential deficiencies in communications between the various disciplines is another critical element for any company. Although a smaller company may have shorter communication lines than larger concerns, the same risks exist. For instance, each of the following scenarios would impact both small and large companies:

- R&D uses a material not designated by the design, or uses a 'slightly amended' design.

- Marketing/Sales casually observe that the use for which the device is being investigated is 'not intended'—at least not as mentioned on the product's labelling.

- The sales force asks that a particular device be customised for a particular doctor in large quantities.

- The Regulatory Department is blissfully unaware of any or all of the aforementioned scenarios, as evidenced by its continuing to register a product that, due to amendations, no longer exists as shown by current company documentation.

The Quality Assurance and Clinical Affairs Departments are not mentioned in this litany. Their tasks are to *prevent* such possibilities from occurring. Clinical strategy development is an active elements of the overall business.

TIP	Make the clinical strategy part of the overall business strategy!

TIP	A company with a compliant quality system *needs* to have a mechanism in place by which any given project is subject to the same type of controls and checks and balances *as any other elements of the quality system*. By this mechanism decisions and activities can be streamlined and reviewed prior to implementation. Any decision whereby a new or major amended design or indication is declared economically, legally, strategically, and/or otherwise desirable, *must* be made within the confines of the checks and balances defined in the quality system.

GLOBAL STRATEGY

Every clinical strategy needs to take into account which clinical investigations are designed to satisfy which specific requirements. Requirements differ markedly amongst the countries and regulatory bodies, even though a global harmonised baseline is emerging. A company, nonetheless, needs to address these differences simultaneously. Meeting a global strategy involves balancing and weighing each of the variances to attain a consensus resulting in a defined global strategy.

From a strategic point of view, putting all of the company's eggs into one basket is justified only if the money saved balances the business risk to be taken. The role of the Marketing Department is one of assessing the desired global market position and the overall intended use of the product.

It is almost always possible to link the company's demand for safety and performance data with the type of study being conducted or needed. A possible solution for tight company budgets might be to have the first stage of a trial satisfy specific regulatory objectives, and later to run the same protocol with an increased number of patients at the same sites to satisfy another set of requirements.

CONCLUSION

Despite the best of intentions, very few studies are designed or conducted with marketing aspects in mind. In any study it is important to define and clearly state study objectives within the clinical strategy. Failure to clearly define the objectives will ensure that none of objectives will fully be met.

The funding of any study has to be detailed early on in the process; once decisions have been made and the trial is underway, it becomes too late. Excuses to the investigators for cutting off funding are seldom possible or preferable. Transferring the budget mid-way through the study is also not conducive to the progress of the study, nor is it an incentive for the maintenance of the relationship with the investigators and staff. Co-funding also rarely works well. Potential or real or anticipated obstacles need to be dealt with up front. The clinical strategy needs to consider all the possible obstacles and potential solutions and contingencies. A clearly defined clinical strategy plan is a benefit to the company—'e Pluribus Unum'.

6

Clinical Trial Project Plan

Herman Pieterse

Profess® Medical Consultancy BV
Heerhugowaard, The Netherlands

Once the company has determined the clinical development strategy for the product and the types of device trials are known, the actual clinical trial project plan has to be designed. In this chapter, the different phases for the performance of a clinical trial will be discussed with respect to the organisation of the tasks and activities. The tasks that have to be performed will be dealt with by various disciplines in the company. A clinical project leader, a study monitor, and, if applicable, personnel for the management of data will all be part of the project team. In smaller companies, the project leader of the study and the study monitor will be the same person. The tasks and activities are described in the quality system (i.e., the set of standard operating procedures [SOPs] that deal with all phases of a clinical study). In these procedures, the responsibilities of all clinical study personnel have been described. A clinical trial project plan can roughly be divided into four phases: planning phase, preparation phase, execution phase, and analysis and reporting phase.

PLANNING PHASE

In the planning phase, the director of research or any other person with responsibility for the clinical evaluation of medical devices will first appoint a clinical project leader. This person should be qualified to perform the job. The first task of the project leader will be to document a study proposal based on the objective established by the director of research. This study proposal will consist of the following elements:

- the rationale for the study and the objective;
- the budget needed to perform the study;
- the manpower needed to execute the study; and
- a time schedule for the preparation, execution, and analysis and reporting phases.

Any company should first make a development plan for each product. From that development plan, the studies to evaluate the clinical safety and performance can be extrapolated. The management committee will determine which of the various disciplines present in the company—Marketing, Research and Development (R&D), Production, Quality Control, and Quality Assurance—will participate in the proposal. The study proposal will be sent to management, and will be reviewed, discussed, amended if applicable, and then approved.

In order to prepare a thorough study proposal, it is recommended that an operational feasibility study be performed. A feasibility study consists of a number of discussions with field experts. These field experts can inform you on the availability of the patients needed for the study. The following information should be obtained:

- name, location, telephone number, and type of investigational centre;

- estimated number of patients (eligible for the study) seen per month;

- estimated number of patients to be entered into the study;

- any competitive studies ongoing in the centre at the present time;

- interest to participate in the study;

- recommendations for other potential investigators; and

- problems to be anticipated in general, those related to the indication.

From the discussion on the proposed inclusion and exclusion criteria, these field experts (i.e., potential investigators) can determine how many patients can be screened for the study at a particular investigational site. A rough recruitment rate can be calculated from these figures (estimated number of evaluable patients). The recruitment rate and the number of sites that will be selected directly determine the duration of time for the execution phase. In that same feasibility study, an inventory of the costs (laboratory, investigator fees, etc.) can be made. The information from the discussion on money will predominantly determine the budget for external costs.

A feasibility study will be performed to investigate if the proposed project can be executed within the set timetable (number of eligible patients to be recruited in qualified investigational centres), within the budget (if already fixed), and to the technical qualifications (qualified resources). The first phase of the feasibility study will consist of the inventory and assessment of available expertise and resources. Based on the results of this phase, a feasibility plan will be made. This is a proposal made by the project leader on how to execute the "feasibility" study, by whom, and by when. The second phase is the execution of the plan. After the completion of the plan, the project leader will report the findings in the study proposal. In phase three, management will determine the feasibility of the study. The primary person responsible for the feasibility study is the project leader, who must be qualified by training and education.

A feasibility study can be executed by the project leader by following a step-by-step approach documented in the feasibility report/checklist. Instructions are given in the left-hand column. The checklist has been designed as a decision tree and a report (see Figure 6.1).

In the internal feasibility study, the project leader will also discuss with the study monitor the standard time for the monitoring of study sites, the amount of paperwork to be performed, and any other activities relevant for this study proposal. Based on those internal discussions, the project leader will be able to calculate the resources needed to perform the job. Once the company has determined the investigator fees that will be paid to the investigator, it is wise not to differentiate between investigators. It is a small world, and everyone talks with everyone. Reimburse each investigator the same amount of money.

The issue of the rationale for the study and the full description of the objective will later be incorporated in the study protocol, but the study protocol should only be designed once management has given a green light for the study proposal. Once the study proposal has been approved, the preparation phase of the clinical study starts.

PREPARATION PHASE

The study proposal has been approved by management, and the project leader can start preparation. Concurrently, the project leader has obtained good insight into the timetable, the available budget, and the resources needed to perform the study. From the feasibility study, he/she also knows which potential investigational centres might participate in the study and what the study protocol should look like. The study protocol should be written in compliance with the SOPs on study

---

Figure 6.1. Feasibility Study Report/Checklist

INSTRUCTIONS

1. Study the available documentation on

    *   indication;

    *   timetable;

    *   complexity study;

    *   design; and

    *   methods.

2. Communication with the management team is essential and advised. The criteria for feasibility should be assessed: external feasibility in relation to internal feasibility and, if already applicable, the economic aspects.

2a. Together with management, a judgement will be made based on the "rating" of the project. A "crash program" feasibility study is a study performed ending with a number of "assumptions" that will be communicated to the client in the project proposition.

3. The completed clinical project database is a database with information on all projects, the names of the employees who were involved, and the outcome, together with an analysis of problems and suggestions for improvement.

3b. Communication with colleagues is an essential step to avoid reinventing the wheel. Information on the recruitment rate, eligibility of patients, indication as such, who the clinical experts are, and how realistic the timetable is should be collected.

4. The company will make a joint decision on the feasibility of the study. If it is decided to follow a crash program, then it is essential to collect all assumptions (to include in the proposal later).

5. The feasibility plan is the basis for a successful feasibility study; give ample time to decide who to contact and what should be asked and how. Two experts is the absolute minimum; if the answers are not consistent and in the same direction, a third expert must be contacted. Contact with the company board of advisors is advised. Questions on eligibility and availability of patients should be asked as well as the proposed recruitment rate, the investigator fee, the methods and procedures, facilities and staff required for the project, and logistics.

Figure 6.1 continued on next page.

Figure 6.1 continued from previous page.

6. Both qualitative and quantitatve information should be provided. Emphasise recruitment and eligibility of patients within the timetable. Costs should be put in writing.

7. The analysis is essential for future project proposals. Updating of the clinical project database will take place based on the outcome of this analysis. Emphasis should be put on the ASSUMPTIONS to be made for this project proposal.

9. The decision will be made by the project leader. If there are any doubts about the feasibility of the project, then management will be consulted.

FEASIBILITY STUDY PROCESS

1. Is information from the development plan sufficient to perform a feasibility study?

   ☐ Yes   ☐ No

   If No, contact management and request more information:

   _____

   _____

   _____

   If Yes, go to 2.

2. Is the timetable for the feasibility study realistic?

   ☐ Yes   ☐ No

   If Yes, define the criteria for feasibility:

   _____

   _____

   _____

   _____

   _____

   If No, contact management and negotiate; convince management of a realistic timetable and go to 2a.

Figure 6.1 continued on next page.

Figure 6.1 continued from previous page.

2a. Does management agree to change the timetable?

☐ Yes ☐ No

If Yes, go to 3.

If No, consult management and decide on further action:

Crash program ☐ Quit the project ☐ → Go to 4.

Reasons for abandoning the project: _____

_____

_____

Outline crash program: _____

_____

_____

_____

_____

Then go to 3.

3. Make inventory of available expertise (INTERN); go to 3a.

3a. Consult the clinical project database where all expertise has been collected from previous studies:

Is expertise available from previous projects?

☐ Yes ☐ No

If No, go to 3b.

If Yes, describe expertise in terms of who did what, how:

_____

_____

_____

_____

Then go to 3b.

Figure 6.1 continued on next page.

Figure 6.1 continued from previous page.

3b.   Consult other expertise within the company.

Is expertise available within the company?

☐ Yes   ☐ No

If No, go to 4.

If Yes, contact the respective person and find out who knows what about the subject:

Outcome of discussion with the expert:

_____

_____

_____

_____

Then go to 4.

4.   Incorporate the findings of the feasibility study in the study proposal and submit the proposal to the management team:

Arguments for decision:

_____

_____

_____

_____

Outcome decision:

Crash program   ☐ → Go to 5.

Quit the project ☐ → END (report to clinical project database)

5.   Make the feasibility plan:

Target group potential investigators:

_____

_____

_____

_____

Figure 6.1 continued on next page.

Figure 6.1 continued from previous page.

Number of experts to be contacted: |__|

Name, address, and telephone number of qualification experts:

1 _____

_____

_____

2 _____

_____

_____

3 _____

_____

_____

4 _____

_____

_____

Questions to be asked:

1 _____

_____

_____

2 _____

_____

_____

3 _____

_____

_____

4 _____

_____

_____

Once the plan has been defined, execute the plan → Go to 6.

Figure 6.1 continued on next page.

Figure 6.1 continued from previous page.

6.   Results of the feasibility study:

1 _____

_____

_____

2 _____

_____

_____

3 _____

_____

_____

Then go to 7.

7.   Analyze, evaluate, and report the outcome:

_____

_____

_____

_____

_____

_____

_____

_____

Then go to 8.

8.   Assess and check the outcome of the feasibility in relation to the criteria:

Outcome of the assessment:

_____

_____

_____

_____

Figure 6.1 continued on next page.

Figure 6.1 continued from previous page.

_____

_____

_____

_____

Then go to 9.

9.   Decide on the feasibility of the project:

Is the project feasible?

☐ Yes   ☐ No

If Yes, continue the project proposal process.

If No, give arguments for the decision:

_____

_____

_____

_____

_____

_____

_____

protocol format and layout. In this SOP, all items that must be discussed and arranged are mentioned. A preparation check-list has been given in Figure 6.2.

Work Breakdown Structure

The first step the project leader should take to organise the study is to draft a detailed work breakdown structure. This is a table with all of the tasks divided per phase (preparation, execution, analysis, and reporting) in the left column. This table will be extended with information on who will perform the task, the target date, the actual date, the reasons for delay, and the interrelation between the tasks (predecessor tasks). Table 6.1 is an example of a work breakdown structure.

Figure 6.2. Preparation Project Checklist

This checklist serves as a guidance and agreement between the project leader and Quality Assurance about which quality systems will be complied with prior to, during, and at the completion of the project. This checklist is part of a dynamic process (i.e., during the planning and preparation of a project, this document will be updated and amended based on discussions with involved parties or due to changes deemed necessary in practice). Please fill out with black ink; copies are sent to Quality Assurance, the Finance department, Management, and other departments.

Discussion No.	Date	Participants	Function	Initial
1.	|_|_|			
2.	|_|_|			
3.	|_|_|			

Figure 6.2 continued on next page.

Figure 6.2 continued from previous page.

Discussion No.	Date	Participants	Function	Initial
4.	|_|_|_|			
5.	|_|_|_|			
6.	|_|_|_|			

Figure 6.2 continued on next page.

Figure 6.2 continued from previous page.

REGULATORY COMPLIANCE

Study will be performed in compliance with

☐ FDA guidelines   ☐ European GCP   ☐ Declaration of Helsinki

PROJECT DESCRIPTION

Project title _____

Project leader _____

Alternate project leader _____

Product _____

Phase development _____

Project awarded   ☐ Yes   ☐ No |_|_|_|_| Anticipated date awarded

If No, describe stage _____

Figure 6.2 continued on next page.

Figure 6.2 continued from previous page.

No. patients planned |_|

No. sites planned |_|

Study design _____

Reference/control _____

Study outline present   ☐ Yes   ☐ No

If yes, attach copy

If No, describe project _____

[objective, endpoints, assessments] _____

_____

_____

Figure 6.2 continued on next page.

Figure 6.2 continued from previous page.

TASKS TO BE PERFORMED

Tasks	By Whom	Tasks	By Whom
Protocol development	_____	CRF development	_____
Project management	_____	Field monitoring	_____
Investigator search	_____	Investigator qualification	_____
Pre-study visits	_____	Initiation visits	_____
Development documents ethics	_____	Update Investigator Brochure	_____
Design monitor manual	_____	Site instruction manual	_____
Study committees	_____	Medication handling	_____
Notify authorities	_____	Project/forms tracking	_____
Central allocation	_____	Codification of adverse events	_____
Visual check CRF	_____	Statistics	_____
Database design	_____	Data editing	_____
Data entry	_____	Closeout visit	_____
Computer error-check	_____	Internal system audits	_____

Figure 6.2 continued on next page.

Figure 6.2 continued from previous page.

Interim analysis _____   Study-specific training _____

On-site audits _____   Study report _____

DATA MANAGEMENT/VALIDATION LEVELS

A. Full validation (all entries for all CRFs)

B. Target validation (100–20 rule)

C. Sample validation (for a sample of all entries/all CRFs)

D. Sample critical validation (for a sample of critical entries for all CRFs)

For B and D, critical items have to be defined.

Critical items _____

_____

_____

_____

Figure 6.2 continued on next page.

Figure 6.2 continued from previous page.

On-Site Monitoring

Frequency ☐ Every 6–8 weeks

☐ After certain # of CRFs: _____

☐ Other: _____

Activities monitoring

Management on-site	☐ Yes	☐ No
Archiving	☐ Yes	☐ No
CRF check	|_| Level	
Source documents verification	|_| Level	
Drug accountability	|_| Level	

Data Management Procedures

Codification

Medication	☐ Yes	☐ No		
Adverse events			☐ Yes	☐ No

Figure 6.2 continued on next page.

Figure 6.2 continued from previous page.

Data Entry

Single data entry ☐ Yes ☐ No   Double data entry ☐ Yes ☐ No

Data Editing

Visual check by medical reviewer and/or study monitor ☐ Yes ☐ No |_| Level

Computer error-check ☐ Yes ☐ No |_| Level

Auditing

Investigator site ☐ Yes ☐ No ☐ 10% ☐ 20% ☐ other: _____

Monitoring ☐ Yes ☐ No ☐ 10% ☐ 20% ☐ other: _____

General comments _____

_____

_____

TIME PLANNING

Anticipated date documents complete for submission to MEC (1st site) |_|_|_|

Anticipated date supplies to site |_|_|_|

Figure 6.2 continued on next page.

Figure 6.2 continued from previous page.

Anticipated date team trained |_|_|

Anticipated start first site |_|_|

Anticipated finish last site |_|_|

RESOURCE/WORKLOAD PLANNING

Department	# of Persons	Total Workload/ Department	Team Members	Member Trained/ Qualified [Y/N]	
Monitoring	|_|	|_|		|_|	Remarks:
Project leader	|_|	|_|		|_|	Remarks:
Project assistance	|_|	|_|		|_|	Remarks:
DE department	|_|	|_|		|_|	Remarks:
DB department	|_|	|_|		|_|	Remarks:
Statistician	|_|	|_|		|_|	Remarks:
___	|_|	|_|		|_|	Remarks:
___	|_|	|_|		|_|	Remarks:

Figure 6.2 continued on next page.

Figure 6.2 continued from previous page.

Critical Tasks (for planning in order of priority) to Date

1. _____   Comment: _____

2. _____   Comment: _____

3. _____   Comment: _____

4. _____   Comment: _____

COMPLIANCE TO SOPS

SOP no.	Title of SOP	Project Specific Training Needed [Y/N]	Remarks
[ ]	_____	[ ]	_____
[ ]	_____	[ ]	_____
[ ]	_____	[ ]	_____
[ ]	_____	[ ]	_____
[ ]	_____	[ ]	_____
[ ]	_____	[ ]	_____
[ ]	_____	[ ]	_____

Figure 6.2 continued on next page.

Figure 6.2 continued from previous page.

SOP no.	Title of SOP	Project Specific Training Needed [Y/N]	Remarks

FORMS TO BE USED IN THE PROJECT

Form Name	Amendment Needed [Y/N]	Remarks

Figure 6.2 continued on next page.

Figure 6.2 continued from previous page.

PLANNING TRAINING

Subject	Participants	Planned Date
_____	_____	|_ _|_ _|_ _|
_____	_____	|_ _|_ _|_ _|
_____	_____	|_ _|_ _|_ _|
_____	_____	|_ _|_ _|_ _|

Figure 6.2 continued on next page.

Figure 6.2 continued from previous page.

PLANNING AUDITS

	Applicable [Y/N]	Planned Date	Comments
Sponsor audit	[ ]	|_|_|_|	
Database validation	[ ]	|_|_|_|	
Error-check validation	[ ]	|_|_|_|	
Data handling audit 1	[ ]	|_|_|_|	
Data handling audit 2	[ ]	|_|_|_|	
On-site audit	[ ]	|_|_|_|	
Monitoring audit 1	[ ]	|_|_|_|	
Monitoring audit 2	[ ]	|_|_|_|	

COMMENTS/DISCUSSION

Table 6.1. Sample Work Breakdown Structure Tasks

Project # _____

Project Leader _____

Title of Project _____

Subject/ Number of Task	Task	Relation Between Tasks	Responsible Person	Date Planned	Date Actual	Reason for Delay	Remarks
Investigation Plan	Make outline study.			|_|_|	|_|_|		
	Make time plan.			|_|_|	|_|_|		
	Draft study protocol.			|_|_|	|_|_|		
	Discuss first draft within company.			|_|_|	|_|_|		
	Discuss first draft with (principal) investigator.			|_|_|	|_|_|		
	Edit protocol to final.			|_|_|	|_|_|		
	Discuss protocol with statistician.			|_|_|	|_|_|		

Table 6.1 continued on next page.

Table 6.1 continued from previous page.

Subject/ Number of Task	Task	Relation Between Tasks	Responsible Person	Date Planned	Date Actual	Reason for Delay	Remarks
	Send final protocol to Quality Assurance for check.			|_|_|	|_|_|		
	Sign protocol and send to investigators for signature.			|_|_|	|_|_|		
Budget	Draft a budget.			|_|_|	|_|_|		
	Discuss the budget with management.			|_|_|	|_|_|		
	Draft an investigator contract and send to investigators.			|_|_|	|_|_|		
	All contracts have been signed and received.			|_|_|	|_|_|		
Informed Consent	Draft patient information and consent form.			|_|_|	|_|_|		
	Arrange for insurance in case of injury to patients.			|_|_|	|_|_|		

Table 6.1 continued on next page.

Table 6.1 continued from previous page.

Subject/ Number of Task	Task	Relation Between Tasks	Responsible Person	Date Planned	Date Actual	Reason for Delay	Remarks
Case Record Forms (CRF)	Design a patient record form.			|_|_|	|_|_|		
	Discuss the CRF with the investigator and experts in the company.			|_|_|	|_|_|		
	Ask statistician to verify CRF.			|_|_|	|_|_|		
	Finalize the CRF.			|_|_|	|_|_|		
	Print the CRF, if applicable.			|_|_|	|_|_|		
Ethics Committee	Collect all documents to be submitted to Ethics Committees.			|_|_|	|_|_|		
	Submit documents to Ethics Committee.			|_|_|	|_|_|		
	Approval of Ethics Committees received, including list of members with function.			|_|_|	|_|_|		

Table 6.1 continued on next page.

Table 6.1 continued from previous page.

Subject/ Number of Task	Task	Relation Between Tasks	Responsible Person	Date Planned	Date Actual	Reason for Delay	Remarks
Investigator	Make selection criteria for investigators.			|_|_|_|	|_|_|_|		
	Select investigators.			|_|_|_|	|_|_|_|		
	Arrange for prestudy visits.			|_|_|_|	|_|_|_|		
	Check facilities and staff of investigator.			|_|_|_|	|_|_|_|		
Documents	Collect updated and signed CVs from investigators.			|_|_|_|	|_|_|_|		
	Collect normal laboratory values, if applicable.			|_|_|_|	|_|_|_|		
	Arrange for an updated Investigator's Brochure.			|_|_|_|	|_|_|_|		
Archive	Design an archive system.			|_|_|_|	|_|_|_|		
	Design an investigator study file to be kept at the site.			|_|_|_|	|_|_|_|		

Table 6.1 continued on next page.

Table 6.1 continued from previous page.

Subject/ Number of Task	Task	Relation Between Tasks	Responsible Person	Date Planned	Date Actual	Reason for Delay	Remarks
Study Manuals	Make investigator manual with information on logistics, procedures, and methods.			|_|_|	|_|_|		
	Make monitor manual on how to validate the data visually during monitoring.			|_|_|	|_|_|		
	Agree on extent of on-site monitoring (frequency).			|_|_|	|_|_|		
	Make data handling manual on how to clean data in the database.			|_|_|	|_|_|		
Analysis of Data	Design, test, validate, and approve database according to specifications.			|_|_|	|_|_|		
	Agree on computer validation checks.			|_|_|	|_|_|		

Table 6.1 continued on next page.

Table 6.1 continued from previous page.

Subject/ Number of Task	Task	Relation Between Tasks	Responsible Person	Date Planned	Date Actual	Reason for Delay	Remarks
Notification Study	Send notification to Competent Authority.			|_|_|	|_|_|		
	Approval received within 60 days.			|_|_|	|_|_|		
	Arrange import licence, if applicable.			|_|_|	|_|_|		
	Import licence arranged.			|_|_|	|_|_|		
Study Devices	Order devices.			|_|_|	|_|_|		
	Order any other study material: _____ _____ _____			|_|_|	|_|_|		
	Pack and label devices according to protocol.			|_|_|	|_|_|		

Table 6.1 continued on next page.

Table 6.1 continued from previous page.

Subject/ Number of Task	Task	Relation Between Tasks	Responsible Person	Date Planned	Date Actual	Reason for Delay	Remarks
	Make randomization list, if applicable.			_ _ _	_ _ _		
	Make decoding envelopes, if applicable.			_ _ _	_ _ _		
	Send devices to investigator, once all essential documents are in-house.			_ _ _	_ _ _		
	Send other study materials to the investigators.			_ _ _	_ _ _		
Instruction Sites/Initiation	Agree on an instruction/ initiation visit with the site.			_ _ _	_ _ _		
	Instruct investigator/staff.			_ _ _	_ _ _		
Monitoring	Agree on frequency and tasks of monitoring.			_ _ _	_ _ _		

Table 6.1 continued on next page.

Table 6.1 continued from previous page.

Subject/ Number of Task	Task	Relation Between Tasks	Responsible Person	Date Planned	Date Actual	Reason for Delay	Remarks
	Agree on extent of source documents verification.			|_|_|	|_|_|		
CRF Tracking	Design tracking system for CRFs and other essential documents.			|_|_|	|_|_|		
	Design a stock control system for relevant materials (e.g., CRFs).			|_|_|	|_|_|		
Data Handling	Decide on data entry procedure.			|_|_|	|_|_|		
	Decide on data clarification procedures.			|_|_|	|_|_|		
	Plan database clean.			|_|_|	|_|_|		
Statistical Analysis	Decide upon which tests to be performed, why, and how in analysis plan.			|_|_|	|_|_|		
	Perform analysis.			|_|_|	|_|_|		

Table 6.1 continued on next page.

Table 6.1 continued from previous page.

Subject/ Number of Task	Task	Relation Between Tasks	Responsible Person	Date Planned	Date Actual	Reason for Delay	Remarks
Reporting	Decide on format, layout, and contents of statistical report.			_ _ _ / _ _ _ / _ _ _	_ _ _ / _ _ _ / _ _ _		
	Decide on format, layout, and contents of medical report.			_ _ _ / _ _ _ / _ _ _	_ _ _ / _ _ _ / _ _ _		
	Decide on publication strategy.						

Task/Budget Monitoring

In order to determine capacity planning for study monitoring, it is essential to provide the monitors with a task/budget in hours list. For a clinical study with four investigational sites, N equals 4 patients per site to be entered in the study, and a monitor visit each month during a seven-month period, a typical task/budget list for each study monitor could look like the example in Table 6.2. The list in Table 6.2 can be adapted for each study and can also serve to keep a balance of the hours invested in each monitoring visit per centre. The objective of this list is to assist the study monitor to stay within the budget for the study.

First Project Team Meeting

Once the project leader has a complete overview of the tasks that must be performed for the study, then he/she will invite the potential team members to a first project meeting. The objectives of this first internal meeting, where all involved parties (statisticians, data entry operators, study monitors, and other in-house functionaries) will be invited, are as follows.

- Present the work breakdown structure (i.e., the order of the tasks and interrelationship).

- Present the project plan (in time and tasks).

- Discuss the internal task/responsibility scheme (i.e., who will do what by when).

- Discuss the budget of the resources in time.

- Discuss pitfalls, drawbacks, dangers, constraints, unclear issues, weak points, and threats.

In order to organise the first internal meeting, the results of the feasibility study should be known, and the outline of the study protocol should be written.

The first internal project team meeting will result in an update of the work breakdown structure, a specification of some of the tasks (including an adaptation of the budget in

Table 6.2. Task/Budget List Monitoring

Task Monitor Name:	No. of Hr/Site	No. of hr Total	Balance After Month							Net Balance	Specifications/Remarks
			1	2	3	4	5	6	7		
N = 4 patients/site; Budget for all sites											
Prestudy visits	7	28									
Site 1	7										
Site 2	7										
Site 3	7										
Site 4	7										
Medical Ethics Commitee approvals	3	12									
Supplies management	1	4									
Investigator file	0.5	2									

Table 6.2 continued on next page.

Table 6.2 continued from previous page.

Task Monitor Name:	No. of Hr/ Site	No. of hr Total	Balance After Month							Net Balance	Specifications/ Remarks
			1	2	3	4	5	6	7		
Initiation visits	8	32									
Site 1	8										
Site 2	8										
Site 3	8										
Site 4	8										
Monitoring visits	8	224									
Site 1	56										
Site 2	56										
Site 3	56										
Site 4	56										
Study closure visits	8	32									

Table 6.2 continued on next page.

Table 6.2 continued from previous page.

Task Monitor Name:		No. of Hr/ Site	No. of hr Total	Balance After Month							Net Balance	Specifications/ Remarks
				1	2	3	4	5	6	7		
	Site 1	8										
	Site 2	8										
	Site 3	8										
	Site 4	8										
Visits (include travel and reporting)												
Internal training		16	16									
TOTAL			350									

hours per task, if applicable) and a complete, detailed activity list for the tasks to be performed in the preparation phase.

Other primary objectives for the first project team meeting are to build team spirit, to have all participants facing in 'the same direction', to agree on the plan of action, and so on. Although everything at this preparatory phase seems feasible, well organised, and well planned, it is essential to plan pitfalls, and to talk about a worst case scenario. Depending on the interpersonal and professional skills of the project leader, the first meeting can be either a success or a failure.

At the first meeting, the frequency of team meetings will be determined. The frequency of team meetings depends entirely on the critical tasks to be performed and/or milestones that have been reached. In the preparatory phase, the team will probably meet more often than during the execution phase of the study.

Operations Manual

In most companies, the clinical project leader will perform many of the tasks oneself. Depending on the complexity of the study, the project leader will lead a smaller or a larger project team. The smaller the team, the more easy the management of the team and the evaluation of the progress of the study. One of the primary tasks after the first project meeting has been held is to finalise the study protocol, to select potential investigators, and to design an Operations Manual.

The Operations Manual is a document that gives instructions to different involved parties; documents the tasks and responsibilities of the different involved parties; and serves as training material for the optimal training and qualification of investigators, research nurses, laboratory personnel, and monitors. The primary objective of the Operations Manual is to provide practical guidance to involved parties in order to prevent mistakes during the execution of the study. The prevention of mistakes and optimal communication will increase the efficiency of the project. The Operations Manual has to be concise (i.e., legibile, user-friendly) and short and should serve its purpose only. If instructions are more effectively given in other documents, then that part of the Operations Manual is not relevant.

The primary person responsible for the design of the Operations Manual is the project leader. Communication with all involved internal parties is essential. Because the project leader is responsible for the proper conduct and efficiency of the study, he/she has to determine the contents, layout, and format of the Operations Manual. The mandatory elements of the Operations Manual are as follows:

- a list of which SOPs (date and version) are applicable for the project;

- clear, concise survey of tasks and responsibilities of involved parties;

- clear, user-friendly instructions, unencumbered by information destined for another party;

- names and addresses of relevant parties for easy retrieval; and

- clear instructions on amendment handling.

The Operations Manual can consist of the following parts:

- investigator handbook (or plastic pocket card),

- laboratory manual; and

- monitor manual (with elements of the data handling manual integrated).

An Operations Manual can be designed by using the checklist given in Figure 6.3. A survey of the tasks and activities to be performed in the preparation phase of a clinical study was given in the project work breakdown structure. The other phases of a clinical study (i.e., the execution phase and the analysis and the reporting phase) should be scheduled in the project plan as well.

In summary, the project plan of a clinical study will consist of a number of documents:

Figure 6.3. Design of an Operations Manual

INSTRUCTIONS

Investigator Manual

Sources for the design of an Operations Manual:

- Protocol
- CRF
- Contract investigator
- Correspondence sponsor
- Prestudy visit reports
- Data handling manual

The investigator should obtain clear instructions on how to fill out all items in the CRF. Always use the first two CRFs as a test for the assessment of the instruction manual.

Use tabular mode. Copy the names and adresses, including telephone numbers, from the protocol. The study worksheet should consist of those key items that normally will not be documented in the patient file.

Laboratory Manual

If the instructions are absolutely clear from other documents like the protocol, then omit the laboratory manual. Once you have made certain elements of the investigator manual, you can use these for this manual too.

Monitor Manual

A monitor manual is THE guidance for uniform conduct of the monitoring process and will ease the data entry and editing process. Code each CRF entry with a colour or other indication as follows:

Yellow: ABCD

Red: ABC

Blue: AB

Mention which item should be consistent with which code. Information on source documents verification should clearly state:

Figure 6.3 continued on next page.

Figure 6.3 continued from previous page.

- What are the source documents?

- What should the contents of the source documents be?

- What entries should be checked with the source?

Start the design of the manuals once the initiation visits have been scheduled. The monitor manual should be finalised prior to monitor training. The monitor manual is the basis for the training of the monitor.

Steps in Manual Design

1. Design the manual in steps. Communicate optimally with involved parties.

2a. Remember that investigators are very busy and do not have the time to read thick manuals.

2b. Give clear, short instructions to the monitor concerning the review of the CRF. Take care that the instructions for in-house review and medical review have been integrated in the monitor manual.

INVESTIGATOR MANUAL

1. Do you consider the instructions in the CRF sufficient to assist the investigator fully?

   ☐ Yes   ☐ No

   If Yes, then concentrate on the following subjects in the investigator manual:

   - Table with tasks for investigator, research nurse, laboratory personnel, monitor, CRC (clinical research co-ordinator)

   - Names and addresses

   - Small plastic card with inclusion and exclusion criteria and adverse event reporting

   - A 'Study Worksheet', to be included in Appendix

   If No, then include the following in the investigator manual:

   - Items mentioned above

   - Instructions on how to fill out the CRF correctly with information on completeness, how to correct mistakes,

Figure 6.3 continued on next page.

Figure 6.3 continued from previous page.

accuracy, and consistency checks. This should preferably be done for each item mentioned in the CRF.

Detailed instructions:

- Make the table with tasks based on the work breakdown structure and protocol. Use the costing sheet.

- Collect names and adresses from the protocol and put them in alphabetical order for better retrieval.

- Include in the manual a survey of the scheduled milestones.

- Make a study worksheet together with the principal investigator.

- Make a pocket-size card with inclusion and exclusion criteria and information on serious adverse event reporting procedure. Laminate the card and distribute it to investigators.

- Define and document which source documents are applicable for this study and what they should contain (see design monitor manual for detailed information).

LABORATORY MANUAL

2. Does your project include complex laboratory procedures and handling?

    ☐ Yes   ☐ No

    If Yes, then design a laboratory manual and make use of the following instructions for the chapters:

- Study flowchart of sample tracking

- Information on who is doing what when

- Details on the laboratory procedures, analysing methods, etc.

- Details on shipping and labelling procedures

- Instructions on which forms to be filled out and the distribution of these forms

Figure 6.3 continued on next page.

Figure 6.3 continued from previous page.

- Include parts of the investigator manual if you deem this relevant for the objective: clear, obvious instructions for laboratory personnel and research nurses

If No, then do not design a laboratory manual.

MONITOR MANUAL

A monitor manual is essential for the proper conduct of the monitoring process. The instructions given for the review of a CRF must be integrated with the instructions for in-house review and medical review. This manual will serve as *the document and guidance* for the training of monitors. A monitor manual should consist of the following chapters:

I.   CRF Review Instructions   Completeness checks (A)

Accuracy checks (B)

Consistency checks (C)

The instructions given to the monitor should be integrated with the instructions given to in-house CRF review and medical review of adverse events, concomitant diseases, and medication.

Give instructions on how the CRF should be checked:

Always check the first CRFs of any centre for 100% of the entries, then:

a.   100% check of 100% of all CRFs

b.   100% check of key variables and a random check of the other entries

The degree of CRF check should be documented and agreed on with the client.

Key items must be defined in case option b is chosen. These are as follows:

- The existence of the patient

- The validity and timing of informed consent

- Inclusion/exclusion criteria

Figure 6.3 continued on next page.

Figure 6.3 continued from previous page.

- The principal assessment criteria
- Adverse events
- Study and concomitant therapy
- Visit dates

II. Source documents checks (D)

Define in detail which entries in the CRF must be checked against source documents and to what extent.

III. Instructions on how to arrange device accountability and medication handling

IV. Instructions on adverse event reporting (role and tasks monitor)

V. Instructions on the contents of the investigator file and the essential documents

VI. Instructions on reporting: which documents, distribution of originals, and copies

VII. Instructions on *trivial* corrections to be made by the study monitor, for example,

- Legibility corrections
- Writing out abbreviations for data entry
- Filling out units, if evident
- Filling out patient No. on top page, if evident

VIII. Frequency of monitoring visits

IX. Tasks to be performed at monitoring visits

X. "Ghost CRF": a fully annotated CRF with remarks for the monitor on how to check data

XI. A recent calendar for planning of monitor visits

STEPS IN MANUAL DESIGN

Responsible person: I(PM)

1. Decide on which manuals serve the objective of optimal instructions.

2a. Draft the manual(s) based on the source documents you have.

Figure 6.3 continued on next page.

Figure 6.3 continued from previous page.

If available, use a manual that already has been made according to these instructions.

2b. Make a ghost CRF with full instructions for the monitors concerning what must be checked for which entry in the CRF.

3. Discuss the draft with relevant parties (investigator manual with investigator, lab manual with a laboratory technician, monitor manual with one of the monitors).

4. Edit the suggestions, ideas, and corrections.

5. Discuss the final draft with the liason of the sponsor.

6. Make the final version and submit it to the sponsor for approval.

7. If a plastic card must be made for the investigators, then arrange this.

- the approved study proposal with information on the objectives, the timetable and budget, and a first estimate on the capacity needed to perform the study;

- a complete work breakdown structure of all tasks to be performed; included in this work breakdown structure is a task/responsibility scheme with information on who will do what by when;

- study-specific documents (e.g., the first draft of the study protocol); and

- an Operations Manual with all relevant information on the project that has not been described in other study documents.

7

Quality Assurance for Good Clinical Practices

Marja G. de Jong
Medidas
Wijk bij Duurstede, The Netherlands

Herman Pieterse
Profess® Medical Consultancy BV
Heerhugowaard, The Netherlands

Good Clinical Practice (GCP) requirements are found throughout various documents governing clinical research[1,2,3,4]. The requirements set forth in these documents are primarily sets of management procedures that ensure that data are correctly recorded and that the rights of patients are protected by independent overview[5].

---

1. U.S. FDA regulations for Investigational Device Exemption (IDE).

2. EN 540, Clinical Investigation of Medical Devices for Human Subjects.

3. Medical Device Directive, 93/42/EEC.

4. Active Implantable Medical Device Directive, 90/384/EEC.

5. Allen, M.E, ed. 1991. *Good Clinical Practice in Europe*. U.K.: Rostrum Publications, p. 7.

Under the GCP requirements, both the clinician and the clinical trial sponsor take on wide-ranging responsibilities. These responsibilities are the first line of a quality assurance system that can be verified by regulatory bodies. The essential element of the quality assurance aspects in GCP is the verification and accuracy of the data to reflect or refute the claims that the manufacturer is making or intending to make concerning the use of the device; that is, the assessment verification of the manufacturer's claims for the technical performance of the device.

In this chapter, the quality assurance aspects of GCP are examined. Emphasis will be on defining the documents that are necessary to manage and run a clinical trial effectively and to ensure the quality of the clinical trial data.

REQUIREMENTS/CRITERIA

Inherent in the GCP requirements are the overall structure of clinical trials and the establishment of a disciplinary approach to ensure an accurate and scientific approach to clinical trials. The primary elements behind GCP requirements are the

- confirmation of compliance concerning the characteristics and performances of the devices under normal conditions;

- determination of any undesirable side effects; and

- assessment of whether there are acceptable risks.

The primary objective of legislation is to

- ensure product safety and efficacy;

- protect subjects enrolled in the clinical investigations;

- establish and ensure the scientific conduct of the investigation; and

- establish a code of conduct to be followed by manufacturers, clinicians, monitors, and agencies involved with the research.

These requirements provide a framework for preparing written procedures for the organisation, design, implementation, data collection, and documentation of the clinical investigation[6]. The major elements as stipulated in GCP can be summarised as follows.

- Ethical conduct (Declaration of Helsinki and its amendments)
- Ethics Committee International Review Board (IRB) review/overview
- Informed consent
- Methodology:
  - Standard Operating Procedures (SOPs)
  - Investigator's Brochure
  - Case Report Forms/recordkeeping
  - Archiving
  - Clinical Investigation Plan
- Responsibilities of the sponsor, monitor, and investigators
- Qualification of the appropriate parties to perform their responsibilities
- Adverse event reporting
- Labelling/handling of devices for the clinical investigation
- Data analysis and results report/critical evaluation
- Compensation in case of injury

These elements need to be covered by and form the basis of the SOPs that govern the clinical trial process.

 TIP Take ample time to design a set of SOPs for the whole clinical research process.

---

6. EN 540, Article 1.3.

ORGANISATION OF THE
CLINICAL INVESTIGATION

To handle clinical investigations, a systematic approach is advantageous. This section defines the elements involved in the organisation of a clinical investigation.

Establish a Project Team or Steering Group

Appointment of a dynamic project leader should be done—a person who is not only dynamic but also knows how to manage and motivate people and how to get things done.

 With any type of group or committee, avoid management by committee. The adage of "When everyone is responsible, no one is responsible is very appropriate".

The composition of the project team should be a representation of the various disciplines involved with the device in question. The responsibilities of the project team should include

- defining and writing the procedures and policy statement;
- implementing the policy and procedures at all organisational levels involved with the trial process;
- strategetic and tactical planning;
- time planning;
- monitoring the process;
- motivating the clinical trial personnel and project members;
- co-ordination of activities;
- allocating resources; and
- budgeting.

Strong commitment from top level management at the internal sponsor organisation and from the external clinical research site is necessary to support the activities and deliverables of this team.

Preparation and Dissemination of a Timeline Chart

A timeline chart can break down into manageable entities all of the elements involved in the organisation of the clinical trial process. It should include the names of the responsible persons for each element, deadlines, milestones, and crucial entities. The chart can be as detailed as possible.

Determine HOW Clinical Investigations Are Handled Within the Company

Define the company's "clinical trial policy", which should be included in the Quality Manual.

TIP	Add a statement concerning the quality aspects of clinical investigations under the section pertaining to "Design Control" or introduce a separate section in the Quality Manual.

Should there be no policy statement in the Quality Manual, establish a clinical trial policy. Policy statements should be salient points that clearly set forth what the company considers a clinical trial to be (i.e., definition), which department(s) are responsible, and how and when clinical trials are to be performed. A sample of a policy statement could be:

> All clinical investigations performed by [COMPANY] are done in accordance with the GCP requirements defined in [NAME OF LEGISLATION] that we are to meet. All documents used in the clinical trial process are subject to control through the appropriate procedures defined within our Quality System.

Some clarification about the type of trials with which the company is involved may be useful as well.

Documentation Requirements

Determining which documentation, as well as the quantity of documentation, is required for the effective management and execution of clinical trials, is one of the most important items that the project team undertakes.

TIP	In preparing clinical trial procedures, always keep in mind that under the European Union's directives, a quality system assessment could include a review of all SOPs and processes pertaining to the clinical trial process.

TIP	Remember that it is important to document precisely how the clinical investigation is performed. Avoid wish lists!

The quantity and the type of documents are determined by the GCP requirements. SOPs need to be available for all aspects of the clinical investigation programme.

TIP	The duration of the design and for the review of SOPs is one year without exception.

Table 7.1 gives an overview of the documents that, as a minimum, are required for GCP compliance.

ORGANISATION OF THE DOCUMENTATION

Writing and maintaining procedures is indeed an arduous task. Procedures are necessary, however, in order to ensure the continuity of quality. They allow quality systems to be non-individual related. By examination and training in the documentation present, any new staff member should know exactly what is to be done, what has happened to date, and what still needs to be done. In preparing documentation, there are some key steps:

Table 7.1. Type of SOPs Needed for a Clinical Investigation

Writing of Controlled Documents	Training
Change Control of Controlled Documents	Handling Non-conformances/Fraud
Issue of Controlled Documents	Handling of Clinical Investigation Devices
Record keeping/Archiving	Statistical Analysis of Data
Clinical Investigation Plan— format, review	Investigator's Brochure—format, review
Final Report Writing Procedure	
Obtaining Ethics Committee Review	Informed Consent—format, review, obtaining
Adverse Event Management	Insurance, Indemnity, Compensation
Initiation Visit/Assessment Visit	Monitoring and Auditing Visits
Contract Review	Competent Authority Submission/Approval

Quality System Examination

Examine the existing Quality System within the company as it pertains to the documentation control aspects. Apply these procedures to all documents written for clinical investigations. Avoid writing new procedures if possible; amend procedures to incorporate the clinical trial aspects. Should this not be possible, then write specific clinical trial procedures. Documentation should cover the key elements mentioned in Table 7.1.

Ensure that procedures exist that define the following elements:

- house style of the procedures;
- change control mechanism and approval/review; and
- issue, retention, and archiving.

TIP	Use procedures already present in the company quality system. Expand their scope to apply to clinical trials. SOPs should be written sufficiently broad to cover all clinicals that the company may perform. There should be no distinction made between different types of clinical trials.

From a practical point of view, several procedures should be written, each covering at least one element (or a maximum of two elements—no more).

TRACEABILITY OF DOCUMENTATION

Ensure that all documentation has sufficient traceability. This is generally applicable to any quality system document. Documents require some very basic elements for traceability purposes. Such information should be incorporated on each page of the SOP as well as on the title page. Should the pages of the procedure become separated, then each page is traceable to the original document.

The following basic elements are to be included:

- date of revision;

- revision level;

- names or initials of persons authorising the release of the documents for use;

- date of implementation or effectivity;

- title of the procedure;

- number of the procedure along with a document prefix, for example, SOP;

- page number and total pages (e.g., Page 1 of [total]); and

- Document History Form or amendment record. This document can be the first page of the procedure and gives an overview of the changes done to the document over time.

Body Layout (House Style)

TIP	The house style of operating procedures and documents under a Quality System needs to be identical for all procedures and documents. Everyone should not be creating their own style for procedures and documents.

Ensure that in the body of the procedure headers and sub-headers are used to convey clearly the information to the user of the procedure. A sample format/layout is provided in Appendix A. Use multi-level numbering.

SOPs should use short, clear, and concise sentences, grammar, and terms. Avoid the use of long, complex terms.

TIP	Remember the target audience for whom you are writing. Looking linguistically intelligent does not impress; in this case, it could lead to complete ineffectivity if the target audience does not understand what is being stated.

Crucial to a good procedure is finding the balance between putting too much information into an SOP and being too vague and unclear so that no one can use the procedure effectively. It is harder to write short documents than long documents. The longer the document, the more detail and the less clear the instruction becomes regarding the procedure to be conveyed to its user/reader.

TIP	The aforementioned statement is crucial. Follow it religiously!

In Appendix B is a matrix of all required SOPs, along with the characteristics and key elements that should be included in the SOPs. The list is not comprehensive. Variations may be applicable that are determined by the product and how a clinical investigation is done by a company.

CONCLUSION/SUMMARY

Writing the procedures necessary to undertake a successful trial is a crucial and time-consuming process. Processes and actions defined in procedures need to be established in the simplest of terms. Avoid whenever possible bureaucratic double checks and controls.

Ensure that each individual has a commitment to and an understanding of the objective of the clinical trial. The latter is crucial: Without commitment and discipline, the procedures will not compensate for the lack of good clinical trial data. The quality assurance aspects that are defined in procedures should lead to a quality trial with accurate and reliable data and results.

APPENDIX A. SAMPLE PROCEDURE

PREFIX: SOP

DOCUMENT NUMBER: 0001

Page 1 of [total #]

TITLE: Writing Controlled Documents

PROPRIETARY & CONFIDENTIAL DOCUMENT

REVISION	AUTHOR OR CHANGE REQUESTOR	IMPL. DATE	REVISION DATE	AUTHORISED BY	DESCRIPTION OF THE RELEASE
00	ZZZ	31-JAN-97	01-Jan-97	XXX YYY	Release

Appendix A continued on next page.

Appendix A continued from previous page.

AUTHORISATIONS FOR RELEASE, SOP-XXXX, REV ##

Name/Function: _____ Signature: _____ Date: _____

Name/Function: _____ Signature: _____ Date: _____

Name/Function: _____ Signature: _____ Date: _____

Appendix A continued on next page.

Appendix A continued from previous page.

PREFIX SOP	DOCUMENT NUMBER 0001	REVISION DATE 01-Jan-97	REVISION 00	IMPLEMENTATION DATE 31-Jan-97
				Page 2 of [Total #]

TITLE: Writing Controlled Documents

1.0 SCOPE

This section defines to what the procedure applies. Include the department or clinical study site, or company or sponsor, whichever is applicable.

2.0 INTRODUCTION OR PURPOSE STATEMENT

This section gives a summary of the procedure along with any relevant information, permitting the user to put the procedure in the context of its use.

3.0 APPENDICES

This is a list of actual forms or recommended forms or flowcharts that are applicable to the procedure being used.

4.0 DEFINITIONS

Define any terms that may need to be defined as applicable to the procedure.

Appendix A continued on next page.

Appendix A continued from previous page.

PREFIX SOP	DOCUMENT NUMBER 0001	REVISION DATE 01-Jan-97	REVISION 00	IMPLEMENTATION DATE 31-Jan-97
TITLE: Writing Controlled Documents				Page 3 of [Total #]

5.0 REFERENCE DOCUMENTS

This is a list of other SOPs which are cross-linked to the procedure being written or to scientific articles.

6.0 RESPONSIBILITIES

Define who is responsible for what and when. List the functions; avoid personnel names. Each function responsible for the activity defined in the scope of the procedure should be included. If responsibilities include those of the investigator, include them. Each SOP needs to have responsibilities included.

7.0 MATERIALS

A list of any materials that may be needed. This section may be optional. In some cases, this section may not apply.

Appendix A continued on next page.

Appendix A continued from previous page.

PREFIX SOP	DOCUMENT NUMBER 0001	REVISION DATE 01-Jan-97	REVISION 00	IMPLEMENTATION DATE 31-Jan-97
				Page 4 of [Total #]

TITLE: Writing Controlled Documents

8.0 PROCEDURE

A chronological, sequential, step-by-step process. Flowcharts can be used to define the procedure to be followed. If verbiage is preferred, keep it concise, short, and in chronological order, from the first step to the last step. This section needs to include how any non-conformances and corrective actions will be handled and implemented. This may be detailed or include a cross-reference to an already existing procedure on non-conformances.

9.0 NON-CONFORMANCES/DISCREPANCIES

This section needs to include the mechanism of how non-conformances or discrepancies are identified, handled, and rectified. Under this section, a statement of recording, implementation, and follow-up of corrective actions is imperative.

PROPRIETARY AND CONFIDENTIAL DOCUMENT

APPENDIX B. CHARACTERISTICS OF PROCEDURES

Procedure	Characteristics to Be Included
Writing Controlled Documents	1. Define the house style and layout of the documents. 2. Who is responsible for final formatting? 3. Include traceability elements to be incorporated on each page. See Appendix A for example.
Change Control: Approval and Review of Controlled Documents[7]	1. Include a matrix with responsible functions for review and approval. 2. Who has editorial change responsibilities? 3. Does the management representative assigned under the ISO requirements sign off on all of these documents? 4. How is the change control process controlled? 5. Who can initiate changes? 6. Who reviews and authorises changes?
Issue Control of Documents	1. Who issues documents? 2. Who receives them? 3. How are they recalled? 4. What documentation is to be completed for issue or recall?

Appendix B continued on next page.

---

7. Change control procedures include review, changes to quality control documents, authorisation, and release of documents. Such procedures can also include changes to product design. However, for the sake of simplicity, consideration should be given to writing a separate procedure for the change control of products.

Appendix B continued from previous page.

Procedure	Characteristics to Be Included
Record Retention/Archiving Procedure	1. Who is responsible for the archive[8]?
TIP: Use a matrix in the procedure to define specific retention periods for the different types of records.	2. How are the records to be completed?
	3. How are corrections made to data entered into the records?
	4. What are the retention periods for the records?
	5. What records are involved[9]?
	6. Where are records to be retained (for example, off-site, in a central archive, or on magnetic tape)?
	7. Retention of originals and copies should be considered.
	8. How are records to be stored?
	9. How are they to be retrieved?
	10. How are records issued?
	11. How are they to be archived?
	12. Who is responsible for archiving?
	13. Which copies should be archived?
	14. Where is archiving done?

Appendix B continued on next page.

8. Making one person or a documentation control department responsible for the archiving and retrieval of the records ensures continued accuracy and traceability of the information. At the time of third-party audits, it is crucial to ensure the accuracy of ALL records.

9. Types of records include Investigator's Brochure, Case Report Forms, Final Report, Informed Consent, Study Protocol or Investigation Plan, Ethics Committee approval, Competent Authority decision, and reviews of contracts/agreements.

Appendix B continued from previous page.

Procedure	Characteristics to Be Included
Contract Review	1. Who reviews?
	2. Time for the review of the contractual documents.
	3. Can the investigation site fulfil its obligations?
	4. Can the sponsor fulfil its obligations?
	5. How are changes/amendments incorporated?
	6. Who are the signatories?
	7. Include a checklist or some other means for reviewers to complete. This will indicate that the contract has been reviewed and understood. Comments hereby are recorded.
Investigator's Brochure	1. Define the layout and format of the brochure.
	2. Define the salient elements of the contents.
	3. Who is responsible for putting the brochure together?
	4. Who is responsible for the review of the brochure?
	5. Changes to the brochure fall within the scope of the change control procedure.
Case Report Forms	1. Define the layout/format or house style of the CRF.
TIP Never forget that a CRF will also be a complete reflection of the study protocol; no more, no less. Therefore, a link needs to exist between the SOP on the study protocol and the CRF.	2. What are the minimum requirements to be included in the CRF?
	3. Who is responsible for the design?
	4. Who is responsible for the review?

Appendix B continued on next page.

Appendix B continued from previous page.

Procedure	Characteristics to Be Included
	5. How are changes controlled prior to printing?
	6. How are changes controlled after printing?
	7. Who trains the investigators in completing the CRF?
	8. How are the CRFs to be completed?
	9. Which elements are included in the CRF[10]?
Investigational Plan/ Study Protocol	1. What are to be the contents?
	2. What are the relevant sections?
	3. What type of information is to be included?
	4. How often is it updated?
	5. What are the sources for the contents?
	6. How are its revisions handled?
Final Report Procedure	1. Define the structure of the final report.
	2. Who is responsible for the writing?
	3. Who is responsible for the review?
	4. Who has final authorisation?
Informed Consent Procedure	1. What are the literature references and where are they obtained?
	2. Establish layout and the sections to be included.
	3. Define ethical considerations that need to be included and undertaken in the study.

10. This will be dependent on the sophistication of the trial.

Appendix B continued on next page.

Appendix B continued from previous page.

Procedure	Characteristics to Be Included
	4. Define witness consent.
	5. How is the study designed in general terms? How are patients selected? Any assessment of the performance of the device?
	6. How are claims determined?
	7. Statistics that will be applied and how these are determined. A statistical consideration of the sample size of subjects to be selected for the study is essential to prove scientific validity.
	8. How control procedures and comparators are established.
Initial Visit Procedure	1. Define investigator selection criteria.
	2. Define site selection criteria for assessment[11].
	3. Which questions are to be asked? Include questionnaire that can be handed out to the monitors for their initial visit.
	4. Include a checklist to be completed after initial visit[12].

Appendix B continued on next page.

---

11. For instance, this may include accrual rate potential, qualifications of the investigators, commitment to the study, experience in previous trials, reputation with regulatory authorities and industry, and so on.

12. The checklist needs to be objective and provide an objective conclusion concerning the clinical site in question, whether the site will be included in the study or not.

Appendix B continued from previous page.

Procedure	Characteristics to Be Included
Ethics Committee/IRB Approval Procedure[13]	1. Information including documentation to be submitted from the manufacturer to the clinician. Include checklist.

TIP — Once a company has worked with several Ethics Committees, one should make a matrix with the following information: Who is the contact person? When does the committee meet in the coming period? What is the deadline for submission? Which documents does the Committee need and how many copies? How will the Committee be informed during the execution and after the study has been completed? Is the committee properly constituted? Is there a standard submission form? What is the administration fee (if any)?	2. Who is responsible for what? The investigator's responsibility in reviewing the Investigator's Brochure and protocol is to ensure that the study is ethical. 3. Write a letter to the Ethics Committee. Include sample. 4. How will the Ethics Committee be updated when changes to protocol or study are anticipated? 5. Include what to report to the Ethics Committee in case of adverse events. 6. Who will be available for the Ethics Committee when necessary? What changes are to be reported to the Ethics Committee and how are they to be updated? 7. What happens if the Ethics Committee or Institutional Review Board rejects or terminates the study? Define a contingency plan. 8. How are contingency plans imparted to the parties involved in the study?

Appendix B continued on next page.

---

13. It may be preferable to define a generic study protocol as a template for all protocols. Such a document can include some obligatory text.

Appendix B continued from previous page.

Procedure	Characteristics to Be Included
Monitoring Visit	1. Define the process to be followed before, during, and after a monitoring visit to the hospital. 2. What checklists are to be used? What criteria are used?
Insurance, Indemnification, Compensation	1. Define the type of insurance (a type of "all risk" insurance or something else). 2. How is insurance requested? 3. What documents must be submitted to the insurance company for such a request? 4. What amounts of indemnity are to be covered? 5. Include any indemnity clauses to be required in any documents pertaining to the trial. 6. Include any trade or health care association guidelines that may be applicable.
Adverse Event Reporting/ Medical Device Vigilance	1. To whom are the complaints reported by the clinicians? 2. Is this person identified in the protocol? 3. To which authority are adverse events to be reported? 4. What is the time frame for reporting to regulatory authorities? 5. Which forms, if any, are to be used? 6. Define the investigation process, follow-up, and closure. 7. Define any corrective actions that are to be taken. 8. How are corrective actions implemented?

> **TIP** Consider including in all forms or records for completion a small section to be completed by the people at the sponsor site involved with the determination, implementation, and follow-up of corrective or preventive actions. This will cut down on the number of forms that need to be completed and reduces paperwork.

Appendix B continued on next page.

Appendix B continued from previous page.

Procedure	Characteristics to Be Included
Handling of Investigational Devices	1. Who is responsible for the release of the devices to the clinical site?
	2. Define any specific labelling that must appear on the devices.
TIP Incorporate all the aspects of handling clinical devices as part of the incoming or shipment of goods. These devices will require the proper "for clinical trial use only" labels.	3. Define all packaging or storage requirements.
	4. Define shipping procedures for the devices.
	5. Define any special instructions for the clinician.
	6. Define servicing procedures of the devices, if applicable, by the sponsor.
Training	1. Who is responsible for the training of monitors and/or clinicians? This may be one person or several people.
TIP Translate the SOPs into a training manual for persons involved in clinical research.	2. How is the training documented? Is any employee training record established?
	3. Are training certificates issued at the end of each training session?
	4. Are these retained in the monitor's personnel file?
	5. Is the clinician training recorded?
	6. Where is the record retained?
	7. On what topics are the monitor and/or clinicians trained?
	8. Define qualifications of the monitors.
	9. What are the qualifications for the clinicians, if any?

Appendix B continued on next page.

Appendix B continued from previous page.

Procedure	Characteristics to Be Included
Labelling	1. How is labelling reviewed?
	2. What languages are involved in translation?
	3. Which labelling elements require translation?
	4. How is the translation and review process handled?
	5. Who reviews the labelling?
	6. Who reviews the translations?
	7. Are translators trained? Is this included in the training procedure?
Study Design Procedure	1. What is the study design?
	2. Who is responsible for drafting it.
TIP — The elements in table 11 should be included.	3. What does it include?
	4. What is the layout?
TIP — Consider if there is to be standardized text or not.	5. What are the contents?
	6. Who reviews the design study?
	7. Where is the input obtained from? By whom? How?
Clinical Investigation Project Plan	1. What is it?
TIP — • Ensure the CIPP is standard for all projects. • Ensure the same items are always covered, including budget, resources, and the like. • Include checklists where appropriate.	2. Who is responsible for it?
	3. What is it to contain?
	4. What is the layout? Format?
	5. When is it to start?

Appendix B continued on next page.

Appendix B continued from previous page.

Procedure	Characteristics to Be Included
Validation of Systems Used for Data	1. How has the statistical programme been validated? By supplier? by manufacturer/user?
▭ TIP Per the Quality System requirements, a validation procedure needs to be in place anyway. Either use the existing procedure or adapt it to meet software needs.	2. Define the process of validation, if necessary,
	3. Who does the validation?
	4. Who is involved with the validation process? Team? Department?
▭ NOTE Use the writing procedures to define the house-style of any written SOPs and reports such as validation reports.	5. What goes into the validation plan? Content? Layout?
	6. I revalidation necessary?

Section III

How to Prepare a Clinical Trial

8

Protocol Development

Peter Duijst

Aortech Europe Ltd.
Almere, The Netherlands

In order to assess the clinical behaviour of a medical device with respect to its performance and safety, it is important that a structured approach be followed (i.e., a clinical study needs to be set up). This will be an essential research instrument for reaching scientifically justified conclusions.

Data will have to be acquired in the same standardised manner for each patient enrolled and studied. If this strict organisational approach **is not followed**, the comparison of patients not only will be difficult, but answering the intentional questions of the evaluation will virtually be impossible. To achieve this level of organisation, a clinical investigation plan or a study protocol needs to be defined, describing in detail how the clinical study will be performed.

NOTE | Frequently, the terms *protocol, study protocol, clinical study design,* and *clinical investigation plan* are used synonymously in clinical research. All describe the background and organisation of a research project in a structured and detailed manner.

An appropriate, well-designed protocol will inform the individual investigator and his/her staff about a number of issues relevant to the execution of the study: the rationale, the aims and objectives, the design and proposed analysis, the methodology, and the conduct of the study. In EN 540, Section 5.3, and in the U.S. Code of Federal Regulations (CFR) Part 21, Section 812.25, the requirements for the investigational plan are given. This chapter will discuss the various steps to take when developing a study protocol according to these requirements.

PRE-INVESTIGATION INFORMATION

Before examining the details for protocol generation, a few remarks need to be made with respect to information that has been collected on the device prior to its clinical testing. In the European Union, this information is compiled in the *Investigator's Brochure* (EN 540, Section 5.2) and in the United States in the *report of prior investigations* (21 CFR 812.27). The background and intention of both reports is different from that for the study protocol. Although a distinct overlap exists between the two, they should be considered as separate entities.

The Investigator's Brochure or report of prior investigations is part of the relevant documentation available on the device prior to the start of a clinical investigation. It holds comprehensive information on the background of the design, in-vitro test data, standards that have been tested, a literature survey, and so on. Since it forms the basis and justification for commencing a clinical investigation, investigators will have access to this information. It may, for instance, be used to provide the Ethics Committee with additional information on the device, if so required. The clinical investigation plan, on the other hand, contains the information necessary for the investigator(s) to perform the clinical investigation.

STARTING WORK ON THE
CLINICAL STUDY PROTOCOL

Prior to producing a study protocol, a clear understanding needs to be obtained on the reasons why the clinical trial is undertaken and the key parameters to be assessed. In synergy with the other parties involved, who may have differing expectations of the clinical trial, the essential issues should be worked out and defined.

A strategy as such will not only increase the understanding of the device by all involved, but ultimately lead to a more accurate wording of the clinical evaluation process. It will be helpful in obtaining a common position with respect to defining the essential questions, which will have to be answered in the trial.

 Prior to writing a protocol, all relevant parties (R&D, Quality Assurance, Regulatory Affairs, Management) should make inputs where necessary. This will help avoid situations where the clinical study has ended and the data collected are either incomplete or incorrect when it comes to assessing whether the device is safe and performs according to its intentions.

 Organise meetings with the various departments and clinicians to explore what they consider of importance with respect to the clinical testing of the device.

It should also be realised that the regulatory requirements imposed on the contents of a study protocol may be different in different parts of the world. For this reason, these need to be considered at the beginning of the protocol design process as well. Table 8.1 outlines the differences between requirements in the European Union and the United States.

WHAT SHOULD BE INCLUDED IN A PROTOCOL?

When writing a clinical protocol, it should be kept in mind that the protocol has to outline and support the process of acquiring data in such a way that data collection can be performed

Table 8.1. Information Included in the Clinical Investigation Plan

EU Requirements Under EN 540	U.S. Requirements Under 21 CFR 812.25
References to relevant literature	Purpose of the investigation: name, intended use, objectives, and duration
Basis and justification for the clinical investigation, objective, and duration of the study	Risk analysis
Title of the project, identification of the device	Description of the device
	Monitoring procedures
Names, qualifications, and addresses of the clinical investigators and other participants, sponsor, monitor, and study locations	Labelling
	Consent material: informed consent and patient information letter
Study design, number of devices and subjects	IRB information: list of all names, locations, and chairpersons
Performance parameters	Other institutions where part of the investigation may be performed
Case Report Forms (CRFs)	
Statistical design description	Additional records and reports that will be maintained on the investigation

efficiently, without jeopardising the quality of the data. When a standard operating procedure (SOP) defining protocol format, development, and review is available in the company from previous studies, this should be used as a template for protocol generation. When this is not the case, the protocol layout to be designed should be worked out into an SOP, which can be used for future protocols.

The protocol has to be written as concisely as possible; must give easy access to the user; and should not be unnecessarily complicated nor difficiult to understand.

A variety of items has to be addressed in a study protocol to create a document that will be in line with the requirements of Good Clinical Practice (GCP). Divided into sections and

NOTE | The study protocol is not only a document written to meet and satisfy regulatory requirements; it is also meant for use in daily clinical practice, essentially by the entire investigator's staff, in the interest of the individual patient and the quality of the data to be collected.

subsections, these items will be presented and discussed in the following paragraphs. The order is not necessarily imperative, but it is meant to be convenient as an overview.

The **title of the project**, which in many cases will be presented in conjunction with the study protocol number, is a useful reference in discussions between the parties involved. Apart from a title and protocol number, a version number should also be given to identify the document during the various stages of its development.

TIP | Do not make the title needlessly long and technical. It should accurately cover the intention of the trial. The recruiting effect should not be underestimated. After all, in many cases, it will be the first information about the trial that an investigator receives.

The **table of contents** is a convenient listing of the various sections in the protocol. When properly laid out, it will increase the protocol's accessibility for those who work with it.

The **introduction** presents the device in general terms, emphasises its intended use, and explains the scientific rationale for the clinical investigation. For instance, by defining the device's performance characteristics in relation to the existing experience in a particular field of medicine, and by discussing relevant literature available, the benefits of the device's clinical application can be pictured.

There will be various ways of pointing out to the investigator why the device holds promising properties that make testing in a clinical setting relevant. It is important to explain the reasons why the project has been set up, preferably by demonstrating the logic behind it, and to emphasise when the

NOTE | The more clear the information on the performance of a device, with a keen eye toward the clinical application, the more a clinician will be interested and motivated to participate in the testing of that device.

clinical application of the investigational device may be beneficial to the patient.

Once the device has been introduced, the study plan may be defined. This section will describe in detail how the actual setup of the trial has been arranged. This part of the protocol may easily become a lengthy and confounding overview of the necessary steps that need to be brought to the investigator's attention. To avoid this and to maintain clarity, the information is best presented in a number of subsections.

The **objectives of the study** will be the first of these subsections and will logically follow the introduction. In concise wording, the aims to be achieved should be specified here. Usually, but not necessarily, this will be related to the performance of the device and the safety of the patient.

The **study design** will be outlined, as well as the number of participating **centres** and the total **duration** of the trial. Theoretically, an impressive number of study design characteristics may be recognised for clinical trials with devices (see chapter 2).

In practice, however, only a limited number of characteristics is used: prospective versus retrospective, randomised versus non-randomised, single centre versus multi-centre, comparative versus non-comparative, long-term versus short-term follow-up.

| NOTE | Frequently, in drug trials, the design characteristic of *blinding* is applied. In trials with devices, however, from a practical point of view, it is virtually impossible to obscure the identity of the device to the investigator. As such, it will always be known to the clinician what particular device has been used with which patient. Unfortunately, this implies that a certain amount of bias will have to be expected in the study. |

Where the study centres are concerned, the complete address of each institution in the study is mentioned, along with the correct names and professional titles of the investigators. The overall duration of the entire project, from the entering of the first patient until the termination of the follow-up for the last patient, will also be mentioned as study duration.

NOTE | The study duration for each patient will be different from the length of the trial, of course, and can be read from the study flowchart or may be deduced from the study plan section in the protocol.

The subsection dealing with **device description** discusses the technical aspects of the device and the scientific principles revealing the background of its intended use. Although the information presented here needs to be fairly detailed, attention should be paid to avoid generating an unnecessarily complicated piece of information, which will not be functional in the practical use of the protocol.

The details presented and discussed should always be judged on their functional relevance for the investigator and be of use in the context of the device's behaviour in the clinical setting. The best way to reach this goal is to use non-confounding information.

TIP | Be careful when using, without clarification, scientific terms that are not necessarily part of the standard jargon of the investigators. Use of scientific terms makes sense only when they undeniably add to the understanding of the device and the protocol.

The **labelling** of the device can be mentioned here as well. According to Annex I of the Medical Device Directive (MDD, 93/42/EEC), the label for the device intended for clinical use shall bear the words *exclusively for clinical investigations;* according to 21 CFR 812, the label shall bear the following information: *CAUTION—Investigational device. Limited by federal (or United States) law to investigational use.*

Attention should also be given to **risk versus benefit** aspects for the patient participating in the study. Since it is expected that the clinical investigation will demonstrate clinical benefits that outweigh the disadvantages of the device, a correct protocol will list not only the expected benefits but also the potential disadvantages.

For practical reasons, the overview presented will strive for completeness without becoming exhaustive. The focus again will be on clinical relevance, rather than on extensively listing the possibilities that may arise—no matter how insignificant to the patient these may be. Important in the understanding of the benefits and risks entailed with the design of the device is that these are potential in character—a matter to be well emphasised in the protocol.

| NOTE | A well-defined listing of the risks and benefits is of assistance not only to the investigator but also to the Ethics Committee or Institutional Review Board (IRB) when assessing patient safety in the proposed investigation. |

| TIP | Once a well-balanced set of potential risks has been formulated, the information will be very useful in the design of the adverse event form. |

Under the **patient selection** heading, the **number of patients** and the **inclusion and exclusion criteria** are presented and discussed. Although a simple statement on its own, the procedure for developing a scientifically justified number of patients (sample size), resulting from a correct statistical design, can be a complicated affair. The considerations and reasoning behind the sample size are usually presented in a separate section of the protocol.

| NOTE | The ultimate number of patients in the study often results from statistical procedures, but in certain cases, it may be derived from regulatory guidance documents (i.e., ISO/DIS 14155, EN 540). |

| TIP | In a multi-centre trial, the total number of patients to be enrolled and the number expected per centre should be pointed out to the investigators when discussing the protocol with them. The chances of ending up with an unbalanced patient distribution in the trial will thereby be reduced. |

The number of subjects and the number of devices should be mentioned. Discrepancies between the two should be explained.

For the process of patient selection, the definition of inclusion and exclusion criteria is essential. On the one hand, these criteria tend to influence subject recruitment, such that the

number of patients that will be allowed to enter the study will
be compromised with pathologies other than those for which
the device is supposed to be the therapeutic or diagnostic
answer. A more or less homogeneous patient cohort will be
created, thus decreasing bias as a result. On the other hand,
the selection criteria should not be too restrictive. Patient
recruitment should permit the entering of a sufficient number
of patients, such that scientifically justified conclusions may
be acquired.

These are the two main factors that, by their fundamentally
opposing characters, complicate the definition of selection criteria. It may be easily perceived that serious implications for
the conduct of a trial may arise from incorrectly defined inclusion and exclusion criteria. Hence, it will be of major importance that these are fully defined with all relevant parties.

The inclusion and exclusion criteria need to stipulate precisely the conditions that will make an individual patient eligible for the clinical trial. Despite the high level of accuracy
desired, the wording has to be concise, and the sentences have
to be as brief as possible. An arrangement that follows the format of a simple listing will be the most convenient and beneficial to the user.

| TIP | Separate from the protocol, a brief overview of the inclusion and exclu-
sion criteria can be of assistance to the investigator and his/her colleagues. A small reference card will allow a quick eligibility check when
seeing potential patients and can be quite helpful in the recruitment
process. |

The part of the protocol dealing with **patient enrollment**
describes the procedure to follow for eligible patients. These
procedures involve explaining the study, obtaining informed
consent, signing the appropiate forms, and assigning the
patient identification number for the study. Information on how
to act in case the patient is unable to give informed consent can
be given here as well. The procedure for patient withdrawal
may also be explained here.

One of the most prominent parts in the protocol is the one
dealing with **ethical considerations.** According to the GCP

| TIP | When the patient identification number used for the hospital files is used in the clinical trial as well, a potential source of confusion is eliminated with regard to data communication between the study monitor and the investigator. |

---

guidelines, the accepted basis for the ethics in a clinical trial is the Declaration of Helsinki. Essentially, the clinical trial ethics are set out as guidelines intending to protect the individual who is subjected to clinical research.

Assurance that the study has been set up in compliance with these guidelines has to be provided to an independent Ethics Committee or IRB prior to starting actual patient entry.

The protocol should mention the ethical considerations and precisely point out what is expected from each investigator: approval of the study protocol by a relevant Ethics Committee or IRB and informed consent given freely by each patient participating in the study. The protocol has to give the procedure for obtaining informed consent and the acceptable deviations from this.

Once the study plans have been established, protocol development may continue with the next section, presenting and explaining the **statistical procedure** that will be applied. The sample size or number of patients to enter the study per centre and the overall number of patients should be mentioned here, in conjunction with the rationale for these choices. In those cases where the study design is randomised, the procedure to follow will be explained in detail as well.

| NOTE | Studies designed to collect data in support of regulatory approval applications and with the aim of publication in medical journals will require a higher degree of statistical analysis. |

The approach for interim and final analysis of the data will be described in the protocol. This may be done in the statistical section or under a separate heading. The parameters that are pivotal to the assessment of the device's performance and safety will need to be determined prior to data collection and need to be defined in the protocol. The criteria according to which the analysis of the various parameters will be performed should also be addressed.

The section dealing with **clinical conduct** will outline what data need to be obtained during the various phases of the collection process. Prior to applying the device, data with respect to the patient's demographics, clinical history, and physical examination will be gathered. These data will allow the establishment of a clinical baseline for the patient group(s) on which the clinical effect of the device may be superimposed. These preprocedural data on the patient group(s) are vitally important when analysing the clinical impact of a device.

The next phase, the **procedural phase**, describes the actual clinical application of the device and the data to be collected. Evidence has to be gathered with respect to the device's justified use in the individual patient. In other words, the findings here should not violate the inclusion criteria, but should confirm the diagnosis made earlier. The device lot number or serial/type number, or any other means of identification to allow traceability of the device afterwards, will be recorded. Clinical findings that may be of relevance to data analysis and complications that may occur during the procedure have to be recorded.

Once the procedure has been performed, the **follow-up phase** is initiated. Over a strictly defined length of time, short-term and, depending on the device, long-term data will be collected. The intention here is to monitor over an extended period of time the clinical effect rendered by the procedure in the individual patient and to accumulate information with respect to the clinical behaviour of the device over time.

| TIP | List the data to be collected in the study for each of the phases, such that they are exactly in line with the design of the CRF. In this way, the CRFs and the protocol form a coherent set, which will help prevent confusion on which data need to be collected. |

In the section dealing with **adverse event reporting**, the procedure for the investigator or data co-ordinator to follow is outlined for those cases in which a severe incident occurs in the trial. In contrast to mild or moderate adverse events, serious adverse events need to be immediately reported to the study monitor. This will be stated in the protocol with the address, telephone number, and fax number of the study monitor or other person.

| NOTE | In the case of disposable devices—for instance, catheters—data collection will be mainly related to the phase of the clinical procedure when the device was applied. Collecting follow-up data will, nevertheless, be important for understanding and evaluating the clinical effects that may be expected over time from the procedure. |

| TIP | Restrict the amount of data to be collected to a minimum. Only data relevant to the study that helps answer the questions raised in the objectives of the study are to be recorded. Non-relevant issues will only burden the parties involved in the trial and reduce motivation. |

| NOTE | Adverse events may be patient-related or device-related. A serious incident with a device involving the user rather than the patient is also considered a severe adverse event. |

What is regarded as a reportable adverse event, should be defined in the protocol. Incidences that are fatal or life threatening are considered serious adverse events. When a disabling effect results from the incident or the adverse event leads to hospitalisation or prolonged hospital stay, the incident is serious in character and requires immmediate reporting (see chapter 23).

| TIP | Stipulate in the protocol that a reportable adverse event not only has to be reported immediately by phone or fax, but also has to be reported on the applicable CRF, like all other adverse events. |

Although monitoring in a clinical study is generally considered a pivotal activity, crucial for the quality of the data collection process, the **monitoring procedure** in many protocols is not given a prominent place or is sometimes entirely neglected. This section of the protocol will outline the different steps to oversee the conduct and progress of the clinical trial. It will define the role of the study monitor in detail. The essentials for each phase in the trial need to be explained, such that the investigator and the staff understand what is expected of them and what they may expect from the study monitor. Since an extensive description of the monitoring procedure can be found in the chapter dealing with this issue, it is only briefly outlined in this chapter.

The importance of the initial contact between the investigator and the study monitor is given; the same is done with respect to the study placement visit. Next, the study initiation visit and the monitoring visit are addressed. Finally, attention is given to the study termination visit and the actual closing of the clinical trial.

TIP | According to the GCP guidelines, the study protocol should mention the name, address, and professional background of at least one of the study monitors. The end of the monitoring section seems a convenient place to provide this information.

The protocol should also mention that the sponsor and the investigator(s) are obliged to **retain the records of the study** (i.e., the data collected, the study protocol, and other study-related documents) for an appropriate time (for specifics, see chapter 22).

When reaching the phase where the study protocol is finalised, two items may not be omitted—a list of **references** and the **annexes** to the protocol. In the reference list, all relevant source documents used in its development will be outlined. Usually, these are publications in scientific or clinical journals, chapters from relevant books, regulatory documents, and internal reports. The references should allow the user of the protocol to trace the source documents, mentioning titles, pages, journal, publisher, and so on.

The annexes consist of one or more documents that may be of use to the investigators and to the Ethics Committee and that may help increase the understanding of the study. Some of the documents that can be enclosed in the annexes include

- a flowchart;
- the Declaration of Helsinki;
- the informed consent form;
- the patient information letter;
- samples of CRFs;
- the label of the device;

- documentation on specific tests applied; and

- documentation on symptom classification systems applied

This overview of annexes is merely an example. The actual contents will reflect the clinical study to be performed.

REFERENCES

Clinical Investigation of Medical Devices. 1994. Draft international standard ISO/DIS 14155.

Clinical Investigation of Medical Devices for Human Subjects. 1993. European standard EN 540.

Closer to Harmonised GCP. 1995. *Applied Clinical Trials* 4(4/5).

Council Directive 90/385/EEC. 1990. Clinical Evaluation, Annex VII. *Official Journal of the European Communities* 189(17).

Council Directive 93/42/EEC. 1993. Clinical Evaluation, Annex X. *Official Journal of the European Communities* 169(1).

Declaration of Helsinki. Recommendations guiding physicians in biomedical research involving human subjects. Adopted by the 18th World Medical Assembly, Helsinki, Finland, June 1964, and amended by the 29th World Medical Assembly, Tokyo, Japan, October 1975; 35th World Medical Assembly, Venice, Italy, October 1983; 41st World Medical Assembly, Hong Kong, September 1989; and the 48th General Assembly, Somerset West, Republic of South Africa, October 1996.

'Good Clinical Practice for Trials on Medicinal Products in the European Community.' 1990. *Pharmacology and Toxicology* 67:361–372.

Guidelines on Good Clinical Practice, Step 4. May 1996. International Conference on Harmonisation (ICH)/DS/EWG E6: Good Clinical Practice, EUGCPCD7/31.

Meinert, C. L. 1986. *Clinical Trials: Design, Conduct, and Analysis*. London: Oxford University Press.

The Scrip Guide to Good Clinical Practice: Documenting Trials to U.S. FDA Requirements. Richmond, Surrey, UK: PJB Publications Ltd.

U.S. Code of Federal Regulations. Food and drugs, Part 21, Section 812.25.

9

The Clinical Investigator's Brochure

Herman Pieterse

Profess® Medical Consultancy BV
Heerhugowaard, The Netherlands

When a device company wishes to initiate clinical studies in the United States or in Europe, a clinical investigator's brochure has to be written. The objective of such an investigator's brochure is to inform the investigator on the properties of the product and the safety tests that have been performed in animals. Also, the performance of the device should have been bench tested.

Both in Europe and in the United States, the confirmation of conformity with the essential safety requirements may require that clinical investigations be carried out under the responsibility of the manufacturer. In the case of implantable and long-term invasive devices in class II-a or II-b and devices in Class III, the confirmation of conformity must be based on clinical data. The performance and characteristics of the devices and the undesirable side effects have to be evaluated under normal conditions and for the intended purposes.

Good Clinical Practice (GCP), particularly as stipulated in EN 540 and ISO 16155, defines procedures that assist manufacturers, regulatory authorities, sponsors, and clinical investigators on the conduct and performance of a clinical investigation of medical devices. The standards not only specify the requirements for the conduct of clinical investigations, but also specify how and what documentation needs to be provided. The investigator together with the staff and the Ethics Committee (or Institutional Review Board [IRB]) shall be provided with documentation outlining the basis and justification for the clinical investigation. A part of this documentation shall be included in a clinical investigator's brochure. For the United States, a Report of Prior Investigations is similar to an Investigator's Brochure. Both documents contain information that has been collected on the device prior to clinical assessment in a clinical setting.

The clinical investigator's brochure is a dynamic document and, due to ongoing findings, has to be updated regularly during the development phase of a product. When submitting this brochure to an Ethics Committee or IRB, it should reflect the current status of development.

| TIP | In general, a yearly update of the clinical investigator's brochure is advised. |

The clinical investigator's brochure is a crucial document during the clinical development phase of the product. The person who is to design such a document should bear in mind that the document represents the actual status of development.

WHICH FORMAT IS PREFERRED?

The format is a matter of company style and culture. The document should be properly addressed and labelled. This means that the footer or header has to contain at least the

- version number;
- date of the version;

- file address; and

- page number (e.g., Page 1 of N).

The format is not obligatory, but it should contain all of the information necessary to give a full review of the results from all technical, biocompatibility, and safety studies. The company should bear in mind that information on prototypes of the product presented could be of importance to ensure that safety has been guaranteed. An example for the contents of a clinical investigator's brochure is given in Table 9.1.

Table 9.1. Format/Contents of a Clinical Investigator's Brochure

Chapter	Subject
	Front cover with distribution list
1	Title page with signature
2	Confidentiality statement
3	Table of contents
4	Introduction
5	Technical information
6	Compliance to standards
7	In-vitro/ex-vivo performance testing
8	Biocompatibility/safety testing
9	Clinical experience
10	Patient insurance
11	Financial and administrative arrangements
12	Compliance to Good Clinical Practice
13	References
14	Appendices

WHAT SHOULD BE INCLUDED
IN EACH CHAPTER?

The following sections give a description of the contents for each chapter of the Investigator's Brochure, referring as necessary to the relevant statements in GCP requirements. Suggestions for a proper format are also given.

Title Page with Signature

The name of the product, followed by 'Investigator's Brochure', the version number and issue date, the signature of the responsible person in the company, the name and address of the company with telephone and fax numbers, and the file address and date of the electronic version have to be included on the title page.

Confidentiality Statement

Stating that the document should be regarded as confidential and meant only for review by an investigator and his/her staff, a potential investigator or consultant, and an applicable Ethics Committee or IRB is recommended. Many companies control the distribution of these documents in order to ensure that this confidential information will not be distributed to third parties (i.e., competitors).

Here is an example of standard phrasing:

> By accepting this document, the involved party agrees that the information contained herein will not be disclosed to others, without written authorisation from the company except to the extent necessary to obtain informed consent from those persons to whom the drug will be administered.

Table of Contents

A proper table of contents will give the reader a quick overview of the contents of the brochure. The numbering of the sections of the document can be according to the decimal system, and headings should follow a certain style. For example,

1. **BOLD AND CAPITALS**

1.1 NO BOLD AND CAPITALS

1.1.1 No Bold and Lowercase

Introduction

The introduction should broadly describe the device under investigation and its clinical purpose. References to prototypes or products from an earlier generation can be made. Furthermore, the objective and purpose of the brochure (to inform both investigators and his/her staff, Ethics Committee or IRB, monitors, and other involved parties on the status of the development of the device) is mentioned in this section.

| TIP | Indicate that the brochure will be updated regularly. |

Technical Information

5.1 Description of the Device

5.2 Drawings of the Components

5.3 Manufacturing Procedures and Quality Control Tests

5.4 Instructions for Use, Installation

5.5 Intended Performance of the Device

The first paragraph of this chapter (5.1) should describe the device and its functional components in general terms: dimensions, materials, its functional components with the rationale of the design, and the scientific concepts on which it is based.

In a second paragraph (5.2), refer to drawings of the components. These drawings could be included in one of the appendices. The drawings will give the reader an idea of the structure and dimensions of the components.

Manufacturing procedures, as far as safety is concerned, complying with Good Manufacturing Practices, or quality system requirements should be described in the next paragraph (5.3) only if this is directly relevant to the understanding of the device. For example, a flowchart of the manufacturing and assembly process could be considered for inclusion. Mention the quality control procedures that were applicable during manufacturing and packaging (e.g., sterility, material strength tests).

In paragraph 5.4, the functions of the device and instructions for use have to be described. If details on the function, instructions for use, and installation are given in detail in the clinical investigation plan or study protocol, then the writer could refer to the investigation plan.

In paragraph 5.5, the intended performance of the device has to be described in detail. 'Intended performance' here refers to future claims with respect to the performance of the device.

Compliance to Standards

During the development of the device, testing takes place in order to assess its conformity with national and international regulations. Such testing may be conducted by third parties (i.e., test houses) or in-house by the manufacturer. A statement should be given referring to the standards to which the device conforms. Include any testing protocols and results in the appendices.

In-Vitro/Ex-Vivo Performance Testing

In this chapter, a compilation of the tests and their results needs to be provided, starting with a description of the design of the device, the materials, quantities, and so on. The in-vitro/ex-vivo/in-vivo performance tests should be described together with a tabulation of the results. It should be demonstrated that the performance of the design is in compliance with the criteria for acceptance. Information on leaching of the materials should be provided. This chapter can be ended with a short discussion on the results and conclusions.

Biocompatibility/Safety Testing

This chapter is a compilation of all of the pre-clinical bio-logical testing, non-clinical laboratory testing, and any other animal studies that have been performed to demonstrate that the product and its components are safe and biocompatible. Biological safety tests are performed to demonstrate bio-compatibility of the materials used. Biological testing should be done to ISO/EN 30993 using all relevant parts of this standard as appropriate.

The tests to be performed should include the following:

- irritation tests;
- sensitisation tests;
- cytotoxicity;
- acute systemic toxicity;
- hemocompatibility/hemolysis;
- pyrogenicity (material-mediated);
- implantation tests;
- mutagenicity (genotoxicity);
- subchronic toxicity;
- chronic toxicity; and
- carcinogenesis bioassay.

The extent of the testing depends on the contact of the device with the human body and the duration of the contact (i.e., risk class).

Clinical Experience

Previous clinical experience, from clinical investigations, published literature, or marketing history, relating to the safety or performance of the device should be mentioned in chapter 9. An outline of the clinical experience to date is also to be given. Any history of the product in countries where the product may be marketed or approved could be described. Parts of the clinical experience that may have been published should be included in an appendix.

Patient Insurance

Information on patient insurance should be provided in the brochure. Ethics Committees or IRBs should be informed that the company has covered its liability (general and risk) with an insurance policy applicable to any damage of the subjects participating in the clinical trial attributable to the intervention or procedure and the investigational product (see chapter 16).

TIP	An insurance policy to cover product liability is insufficient for products that have not yet been put on the market. A clinical study with a device still under clinical evaluation *has* no permit for the market (no certification or market approval. This means that a specific risk liability for the experiments to be performed should be covered by a specific insurance policy.

In order to provide compensation for bodily harm and/or death directly caused by the use of the investigational device within the protocol of the study, as mutually agreed on and as duly signed by all parties involved, the manufacturer needs to state that it has insured all patients (or dependent persons thereof) participating in the clinical trials and evaluating the device.

The insurance policy should cover the following.

1. Damage suffered by a patient as a result of bodily harm inflicted on the insured patient, which on medical grounds can reasonably be attributed to

   a. the device and/or means used in the study; or

   b. handling or manipulations of the insured patient's body within the scope of the study

   provided that,

   - the study was started during the course of the insurance policy; and

- the damage to the health of the insured patient occurred after termination of the study in which the patient took part, and was reported to the insurance company within an allotted period of time.

2. For point 1 above to be applicable, it is important that the damage to the health of the patient is to have developed at the time the patient seeks medical advice regarding symptoms that at that or a later time appear to be the result of this damage.

The policy does not, however, cover bodily harm and/or death that is attributable to negligence, wrongful act, or failure to act on the part of the investigator(s) or the institution(s) where the studies will be performed, or its employees or agents. In order to cover these risks, the investigator should also have an insurance policy covering professional liability.

A copy of the certificate of insurance should be given to all investigational centres. A letter containing the aforementioned contents will have to be prepared for each trial, stating the names of the investigators, the institutions involved, and the number of the protocol agreed on by all parties. A copy of the study protocol (clinical investigation plan) will be attached to the letter, and a copy of the letter will have to be signed by the principal investigator and, if so required by the management of the involved institution(s), by a person on behalf of that institution.

In countries where required by law, the company will have a separate insurance policy to cover liability in clinical studies. The principal investigator will receive two copies of the policy of that insurance and will have to return one copy after signing it, as confirmation of compliance with the requirements established by law and in the policy regarding these insurances.

TIP | In the subject's informed consent/information, a sentence should be added to inform them that a specific insurance policy has been taken out by the manufacturer to cover any bodily harm or injury due to the device under investigation or the experiment as such.

Financial and Administrative Arrangements

Some information on the costs of the trial should be provided to the Ethics Committee or IRB. In most studies, the study devices will be made available to the investigators free of charge. If this is the case, it could be mentioned that non-routine studies prescribed by the protocol will be paid by the sponsor.

The Ethics Committee or IRB could also ask for information about the amount and method of payment and compensation made available to subjects, in order to ensure that there are no problems of coercion or undue influence exerted on the trial subjects. Payments to a subject should be prorated and not wholly contingent on completion of the trial by the subject. If payments are made to the subjects, they should be documented (see chapter 11).

Before the start of the study, a budget for the costs will be agreed to between the sponsor and the investigator. The sponsor will not be responsible for the payment of charges of the principal investigator's hospital for services that have not been agreed on in advance. The sponsor will bear the cost of the required liability insurance (see chapter 16).

All other costs must be contractually agreed to before the start of the study (see chapter 11). The supplied device is to be used only within the scope of the agreed to protocol.

Compliance to Good Clinical Practice

A statement affirming that the device complies with relevant legal requirements apart from those requirements that the clinical investigation is intended to fulfil and that every reasonable precaution has been taken to protect the health and safety of subjects should be added.

The company, the institutions where the clinical studies take place, and the investigators must comply with any prevailing GCP regulations, guidelines, and protocols. GPC rules have to be followed in order to comply with the directives and other prevailing bioethics regulations, as appropriate. A copy of, for example, EN 540 or the Declaration of Helsinki, may be attached to the clinical investigator's brochure. An example of standard phrasing could be:

> Both the investigators and the company will take all necessary precautions to protect the health and safety of subjects taking part in the clinical studies according to the rules as laid down in the revised version of the Declaration of Helsinki [1996 South Africa].

National legal requirements with respect to clinical trial approval, approval by medical Ethics Committees or IRBs, and informed consent procedures need to be obeyed.

Every study requires the permission of the Ethics Committee or IRB that operates in or for the institution in which the clinical trial is to take place. For every patient, **informed consent** (written or verbal witnessed) is to be obtained. The principal investigator accepts the responsibility of writing the information to be given to possible subjects (in accordance with the current Declaration of Helsinki) and of ensuring that each investigator obtains informed consent in accordance with the rules that were established by the principal investigator. The procedure of obtaining and documenting the informed consent must be reviewed and accepted by the Ethics Committee or IRB. The study must be performed according to a detailed protocol. This protocol is to be designed by the sponsor in cooperation with the principal investigator. No patient studies will be performed outside the criteria and procedures in the protocol. The protocol must be signed by both the principal investigator and the sponsor.

The sponsor monitors the progress of the study once every two months at a minimum. Non-compliance with the protocol could result in termination of the clinical trial by the sponsor. The investigator may wish to publish the findings of the study in a scientific journal or present them at a scientific meeting. The company reserves the right, however, to review any publication prior to submission or any paper before it is presented. The results of a multi-centre study may be published as a whole.

References

A complete list of literature references has to be given.

Appendices

The appendices could consist of

- a copy of the relevant requirements (i.e., sections of the Food and Drug Administration (FDA) Code of Federal Regulations, or EN 540, or ISO 14155, etc.);

- the Declaration of Helsinki;

- drawings of the components;

- flowcharts of manufacturing procedures;

- any testing certificates from testing houses or Notified Bodies (in case of commercial products in trial, NB-issued certifications or FDA-clearance or approval); and

- the insurance certificate.

HOW TO DESIGN THE BROCHURE

The following actions should be undertaken.

1. Collect all documents with respect to chapters 5–9.

2. Discuss with management chapters 10 and 11.

3. Make a first outline (headings, selection of the documents).

4. Discuss with a selected group of involved persons (R&D, Manufacturing, Clinical department) the outline, and decide on its contents (which documents to be included).

5. Make a first draft of the brochure (include questions and remarks in the text where the items are not yet evident).

6. Send the first draft to the project group, plan a meeting, and request comments two days before the meeting.

7. Discuss the first draft with the project group.

8. Edit the draft based on the comments from the group.

9. Send the next draft to the group for approval.

10. Steps 8 and 9 may be repeated depending on the comments.

11. Finalize the brochure; the document will be signed and dated by the responsible manager.

With respect to page layout, font type, and font size, choose those fonts and margins that will improve the readability of the document. Each company will have its own style and layout. Footers and headers can be used for pagination, version number, file address, and issue date of the brochure. A front cover sheet, including a distribution list, should give information on the reviewers and other addressees.

10

Budgeting: Monetary and Resources

Philippe Auclair
Guidant S.A.
Zaventem, Belgium

Various steps are required when carrying out a clinical investigation of medical devices. From the initial design of the protocol to the drafting of the final study report, each step has to be followed. One of these steps is often overlooked, namely, "how to assess and define the financial and human resources necessary to complete the investigation successfully". This is a crucial step; any shortage of resources after the investigation has started may result in shortcomings and could ultimately jeopardise the validity of the study.

INFLUENCE OF STUDY PARAMETERS

When considering the resources needed for a given clinical investigation, the following questions—amongst others—are to be answered.

Project Phase

Does the study involve routine medical practice procedures or high-tech procedures? It is clear that the type of study influences the necessary financial resources. A preliminary study involving a handful of prototypes may be more demanding than a study aimed at confirming the long-term efficiency of a device. The methods used for the assessment of efficacy variables may then be highly sophisticated and totally different from routine medical practice.

Complexity of the Study

What is the complexity of the study? A one-month investigation aimed at estimating the sensitisation of a dressing material obviously bears little similarity with a long-term study for an implantable defibrillator requiring high-tech, expensive methods and long-term follow-up.

Duration of Treatment

What is the duration of treatment? The following factors are to be taken into account: the number of visits, the duration of follow-up, and the availability of the hospital team. An investigation on an implantable device could require 20 years of follow-up in comparison to short-term tolerability studies.

Required Staff and Equipment

Is unusual equipment required for the evaluation? Does the procedure require standardisation between centres? If so, is a co-ordinator needed? Is the equipment available at the investigational site or must it be leased or bought?

Nature of the Device

On what class of device is the investigation carried out? Is it being carried out, for example, on a Class I device or a Class III?

Country of Location

Are there any investigator's fees? If there are, the socioeconomic status and legal national requirements need to be considered. In some countries investigators may not receive fees.

Number of Subjects and Centres

How many centres are necessary? Fees are usually calculated on the basis of the number of patients (e.g., the number of completed Case Report Forms [CRFs]).

HOW TO ESTIMATE THE NECESSARY BUDGET

Once these basic questions are answered, the costs can be broken down into the following categories (see Table 10.1).

1. Out-of-house expenses: the investigators' team fees, as well as the equipment, laboratory testing, and all costs related to the team based at the investigational centre(s) carrying out the study.

2. In-house expenses:

   - All costs associated with the clinical trial preparation (e.g., costs for the selection of the monitor/CRA, for the qualification and training of the investigator and the study site personnel).

   - Costs associated with the monitoring activities: How often does a centre need to be visited for monitoring purposes? The assessment needs to take into account the

     - complexity of the CRFs;

     - on-site time needed for the review, as well as checking of source documents, arranging for device accountability, and controlling the documentation; and

---

Table 10.1. How to Estimate the Necessary Budget

OUT-OF-HOUSE EXPENSES

- Professional staff (investigator fees)
- Equipment
- Laboratory services
- Administration
- Contingencies

CONTRACT SERVICES (if relevant)

- Statistical analysis
- Subcontracted work

IN-HOUSE EXPENSES

- Clinical trial preparation
- Monitoring/overhead/travel
- Samples

OTHER

- travel costs, taking into account the number of centres as well as the geographical location of the clinical monitor(s).

3. Contract service(s): Part of a study (e.g., monitoring work) or the entire study can be subcontracted to a clinical research organization.

4. Data analysis: costs for data retrieval, entry statistical analysis, and reporting. These costs can be in-house or out-of-house when subcontracted.

5. All other costs that can be research specific.

For the in-house expenses, it is advisable to have an in-house project team determining the involved costs. Once the tasks and activities have been allocated, then an estimate can be made of the capacity needed to execute these tasks.

Table 10.2 is a tentative summary that can be used as a preliminary budget assessment. Table 10.3 is a non-exhaustive list of the main preparatory tasks to be completed before the study begins.

Table 10.2. Sample Budget for a Clinical Trial

Clinical Trial: _____ Code: _____

OUT-OF-HOUSE EXPENSES

Investigator Fees	Cost per Subject	No. of Subjects	Total
_____	_____	_____	_____
_____	_____	_____	_____
_____	_____	_____	_____
_____	_____	_____	_____
		Subtotal	_____

CONTRACT SERVICES

_____	_____	_____	_____
_____	_____	_____	_____
_____	_____	_____	_____
_____	_____	_____	_____
		Subtotal	_____

IN-HOUSE EXPENSES

_____	_____	_____	_____
_____	_____	_____	_____
_____	_____	_____	_____
_____	_____	_____	_____
		Subtotal	_____

Total Costs for the Clinical Trial: _____

Evaluable Subjects: _____

Table 10.3. Matrix of Preparatory Tasks

Tasks to Be Discussed, Quoted, and Allocated by the In-House Project Team	Budgeting	Responsibilities
Selection of investigators		
Pre-initiation visit		
CRF design		
CRF printing		
Provision for statistical analysis		
Estimate of secretarial/ administrative work		
Laboratory service		
Monitoring:　　　Time needed		
Manpower required		
Travel expenses		
Training of the selected investigator		
Cost of samples, if relevant		

NEGOTIATION WITH THE INVESTIGATOR(S)

Agreeing on a clinical investigation fee with each centre's principal investigator must be a win-win situation resulting from an intelligent agreement. The sponsor should be objective and fair in this negotiation if he/she expects fairness in return. As a basic rule, one ought to be objective, easy on people—but hard on facts.

Negotiating for a clinical investigation is not different from any other types of negotiation. Clinical investigators are just like the people one meets every day. A few can be greedy, whereas others can understate the amount of work necessary to conduct an investigation.

Whereas greediness is to be avoided, let it be said that an investigator who agrees to carry out an investigation for a lower than expected amount of money is even more dangerous for the outcome of the study. Every serious work deserves to be rewarded in proportion to the amount of time and effort spent. The fees are to be sensible. An extraordinary payment that clearly exceeds the amount of work done can be considered as unethical and should be challenged as being an inappropriate incentive from the sponsor. In several European countries, Ethics Committees ask for a copy of the investigator's agreement to check that no undue incentive is given to the investigator; the same could apply to an Institutional Review Board (IRB) in the US.

PROCESS OF NEGOTIATION

In every financial negotiation, the following pattern is used:

- Estimate the cost and the amount of work involved.
- Discuss.
- Deal.
- Action.

The very same steps are used when negotiating with a clinical investigator. The difference lies elsewhere. When you buy a car, you make an offer, negotiate, make a deal, and drive off with it. When negotiating with an investigator, you make an offer, negotiate, and make a deal, but then the clinical investigation begins and everything is up for grabs. If the deal was not a realistic one or the investigator later realises that he/she cannot conduct the study properly with the allocated budget, the whole study is in jeopardy.

CASE STUDY

A typical negotiating situation is given in the following paragraph. Read it carefully and try to answer the three questions raised. Possible answers are also given to help you make your own assessment of this situation.

A monitor, in his/her eagerness to work with a particular investigator, promises to give the investigator future studies if the investigator conducts this first study for a lower price.

Questions:

1. Is there a problem with the monitor's offer in promising more studies?

2. What are the main problems?

3. What are the possible outcomes resulting from the investigator's agreeing to do the study for less money than in the original agreement?

An assessment of the case study:

- If the monitor cannot deliver on his/her promise, credibility will be lost with the investigator.

- The monitor's commitment should not be so definite. There is no way that he/she can be so certain about further studies.

- If the investigator agrees to do the study for less money, he/she might actually lose money, become disappointed, and complete the study without spending the necessary time. The whole study can thus be jeopardised.

CONCLUSION

Assessing the financial and human resources required to carry out a clinical investigation is very important. A less than adequate initial estimation can become a roadblock and can result, if additional funding is not made available, in the failure of the study being successfully completed. Each step of the clinical investigation is to be budgeted, one of the more important steps being the monitoring phase and all related expenses. When negotiating fees with a clinical investigator, the monitor should use common sense and try to achieve a win–win situation that will produce a successful outcome.

11

The Investigator Contract

Philippe Auclair
Guidant S.A.
Zaventem, Belgium

Herman Pieterse
Profess® Medical Consultancy BV
Heerhugowaard, The Netherlands

THE RATIONALE OF AN INVESTIGATOR CONTRACT

The investigator's contract is a financial contract that sets out the investigator's fees and payment schedules. This document should be referred to in case of dispute.

The work carried out by an investigator in conducting a clinical trial should attract a realistic payment in keeping with market forces operating in the applicable branch of the medical profession at the time of the investigation and in the country where the investigation is taking place. Part of the trial planning process involves the project manager, who should commission preliminary surveys of likely costs amongst appropriate professionals in order to plan clinical programme budgets. Because of the different structure and financial organisation of the various bodies that collaborate in research programs, it is not possible to produce a single formula for each financial agreement. For example, in some circumstances, clinical research can be conducted only if extra staff or special equipment is made available beforehand, and the company may be required to fund such staff or equipment in advance of the start of the trial. In other cases, and preferably, a simple "per record form" remuneration will be agreed on, with pro-rated payments for late forms. However, it is always possible to document precisely the financial arrangements for each trial with every investigator, and the key elements of the financial agreement are given in the following sections. They may be incorporated into the study protocol, or they may be part of a financial contract that will be used for longer or more complex studies.

PAYMENT CHEQUES

When a study consists of several sections or when investigators require interim payments, the financial contract specifies in greater detail the payments and the timing of these payments.

Under certain circumstances, for example for studies that involve more than minimal preparatory expense, the total study budget may be paid in units (e.g., thirds): one-third at the start of the study, one-third on receipt of half of the Case Report Forms (CRFs), and one-third on satisfactory completion of the study (i.e., receipt of all case records and missing data requests, along with the investigator's final report). Larger-value studies should be paid in smaller segments—quarters or fifths. Studies paid per patient should be paid at reasonable time intervals. Requests for changes to the name of the payee must be made in writing from the original investigator. Payment in-kind is not allowed.

Requests for payment are made using a special form. Once a cheque is raised by the Accounting department, it is forwarded to the project manager for transmission to the investigator. Accounting will record the payment on the clinical trial financial management system, which forms part of the total clinical trial management system.

PARTICULARS OF THE AGREEMENT

In addition to being a financial agreement, the investigator's contract is also a business document. It defines the scope of the collaboration between the sponsor, the monitor, and the investigator. It includes the precise description of the work to be achieved as well as the responsibility of the different players. All agreements should be signed by the relevant parties and comply with the provisions of Good Clinical Practice (GCP) (see Table 11.1). It also needs to be in compliance with European or local regulatory requirements.

Table 11.1. Responsibilities of Parties under GCP

Tasks and Responsibilities of the SPONSOR

- Investigator selection: criteria.
- Prepare, assemble, and maintain all pre-clinical data and documentation.
- Ensure collection, storage, security, and completion of documents by relevant parties.
- Provide clinical Investigator's Brochure or report of prior investigations and other relevant information.
- Sign the clinical investigation plan (protocol).
- Supply device appropriate to the plan.
- Ensure that provisions are made for compensation in the event of injury.
- Ensure appropriate information and training is given to the investigators.

Table 11.1 continued on next page.

Table 11.1 continued from previous page.

- Do not offer improper inducement to monitor, investigator, or assistants.

- Keep records of any adverse event and adverse device effect that has been reported.

- Agree on recording and analysing method of adverse events and adverse device effects.

- Inform all investigators in writing about serious adverse events and adverse device effects within 10 days.

- Consider jointly with investigators the adverse events and adverse device effects and report them to the authorities.

- Terminate the study and inform investigators.

Tasks and Responsibilities of the MONITOR

The monitor checks and confirms that . . .

- the investigator is informed about the investigational status of the device;

- the investigator knows the requirements to verify the performance of the device.

- periodical compliance with the protocol is maintained;

- any deviation from the protocol is discussed, agreed with, and reported to the sponsor;

- each investigator has the staff and facilities to conduct the study safely and effectively;

- each investigator has continued access to a sufficient number of subjects;

- procedures exist for recording adverse events and adverse device effects;

- the device is used according to documented instructions;

- if modifications of instructions are needed to the device or the protocol, these are reported.

- completion of the documentation is maintained;

- an adequate supply of devices is maintained at the site;

- informed consent is obtained;

- withdrawal and/or non-compliance is being documented;

- data in the CRF conform to subject files; and

- any reason for the termination of the study is documented;

Table 11.1 continued on next page.

Table 11.1 continued from previous page.

Tasks and Responsibilities of the INVESTIGATOR

- Ask Sponsor for information as described in brochure or any other documentation.
- The investigator must be acquainted with the use of the device.
- The investigator must be acquainted with the protocol before signing it.
- The investigator and his/her team must be available to conduct the investigation.
- The investigator must ensure that other concurrent studies will not give rise to a conflict of interest or interference.
- The investigator is responsible for the personal safety and well-being of the subjects.
- The investigator makes the necessary arrangements (e.g., emergency treatment) to ensure proper conduct of the study.
- The investigator endeavours to ensure adequate recruitment rate of subjects.
- If appropriate, subjects will be provided with means of identification as they take part in a study (e.g., contact addresses, marking in medical records).
- In order to ascertain state of health of the subject prior to the study, the investigator examines subjects without delivering any direct therapeutic benefit to them.
- The investigator invites subjects to confirm by a signed declaration that they have disclosed all matters concerning their health and records any current medication being taken by the patient.
- The investigator ensures that adequate information is given to the subjects both in oral and written form on the nature of the study.
- Information to the subject shall be easily understandable: aims, expected benefits/risks/inconveniences, explanation of alternative methods, consequences of withdrawal.
- Payment or any other inducement with no direct benefit for the subject will only be for expense, time, and inconvenience.
- The subject must be made aware of procedures for compensation and treatment in case of injury/disability as a result of participating in the study.
- The investigator gives the subject the opportunity to inquire about details of the investigation.
- The subject is free to refuse or withdraw from the investigation without any sanction.

Table 11.1 continued on next page.

Table 11.1 continued from previous page.

- The investigator gives the subject sufficient time to decide on participation.

- The investigator informs the subject that participation will be kept confidential.

- The investigator informs the subject that the data may be made available to third parties while maintaining anonymity.

- The investigator informs any subject who wishes to withdraw of the consequences of withdrawing.

- Obtain informed consent, preferably in writing.

- Following national policy, documentation of consent is done by dated signature of the subject or by an independent witness, who records the subject's assent.

- Documents show how informed consent is obtained and recorded in emergency circumstances.

- In case no signature of either subject or witness is possible, then document each case to the Ethics Committee or IRB and Sponsor with the reason.

- The investigator submits the clinical investigation plan (protocol) to the Ethics Committee for their opinion and approval; reports to the Sponsor.

- If not included in the investigational plan, information is also given to the Ethics Committee or IRB.

- The investigator submits any significant change of the protocol to the Ethics Committee or IRB for its opinion or approval.

- The investigator informs the Ethics Committee or IRB of any severe or serious device effect.

- The investigator informs the Sponsor and the Monitor without delay about any severe and serious adverse event, and about all adverse device effects—and provisions made to counter them.

- The investigator has primary responsibility for the accuracy, legibility, and security of data, documents, and patient records both during and after the study.

- The investigator signs the CRF.

- The investigator signs and dates any alteration of the raw data and retains the original entry for comparison.

- The investigator discusses with the Sponsor any question regarding the modification of the protocol and shall obtain written agreement.

- In emergency situations, the investigator shall exercise his/her judgement to safeguard the subject's interest. In such cases, prior agree-

Table 11.1 continued on next page.

Table 11.1 continued from previous page.

ment is not feasible and therefore these deviations are not considered a breach of the agreement, but are reported to the Sponsor.

- The investigator ensures adherence to the protocol by all members of the investigation team.
- The investigator records any significant deviation from the protocol.
- The investigator specifies and documents the procedure for recording adverse events and adverse device effects.
- The investigator specifies and documents the procedure for reporting all adverse device effects to the Sponsor.
- The investigator is responsible for the supervision and assignment of duties to members of the investigation team.
- The investigator is responsible for maintaining confidentiality.
- The investigator will retain both clinical records and data for an appropriate period of time.
- The investigator will not release the identity of any subject without the subject's prior consent.

The clinical trial agreement may be either a separate document or included in the clinical trial protocol as signed by the physician. The latter allows a single document to be provided to the investigator, but the scientific protocol and financial details cannot be separated. This can become a drawback if the investigation protocol is forwarded to a third party.

CONTENTS OF THE CONTRACT

Article Contents

I   Objective

II   Organisation

III   Investigator Responsibilities, Records, and Reports

IV   Property/Publication

V   Ownership of Confidential Information

VI   Indemnification and Reimbursement

SAMPLE AGREEMENT: CONTENTS AND TIPS

Below is a sample agreement defining information to be included in the document.

The Agreement

Whereas

- The SPONSOR and INVESTIGATOR agree to carry out a . . . study entitled . . . (hereafter called the STUDY).

- The STUDY will be performed with the DEVICE from the SPONSOR.

- The STUDY will be performed in accordance with the Declaration of Helsinki revised 1975 (Tokyo), and amended 1983 (Venice) and 1989 (Hong Kong) and 1996 (South Africa).

- The protocol for the STUDY (hereafter called the PROTOCOL) has been given to the INVESTIGATOR, and the INVESTIGATOR acknowledges having read the PROTOCOL and any documents appended to the PROTOCOL.

- The INVESTIGATOR warrants that he/she has the authority to sign this agreement.

- For the purposes of this Agreement, references to INSTITUTION and INVESTIGATOR shall refer to INSTITUTION and INVESTIGATOR jointly and severally.

- Individuals (hereafter called the MONITOR) designated from time to time by the SPONSOR will oversee and audit the STUDY and ensure adherence to the PROTOCOL.

In some countries, an investigator is not allowed to accept fees; or if he/she does accept fees, he/she is heavily taxed. This needs to be checked before the study begins.

It has been agreed that:

Article 1: Objective

Under the conditions of this Agreement, the INVESTIGATOR agrees to carry out the STUDY according to the PROTOCOL and to transmit the results to the SPONSOR.

To avoid any confusion, reference the number of the clinical investigation protocol and a statement from the investigator that he/she understands the protocol and agrees to abide by its terms.

Article 2: Organisation

Include such aspects as the following:

- Sponsor shall supply devices, CRFs;

- INVESTIGATOR shall submit PROTOCOL to Ethics Committee for review;

- an agreement on the conditions of the study;

- an agreement on the informed consent procedure to be used;

- an agreement on monitoring procedures; and

- an agreement on non-device materials procedures.

Article 3: Investigator Responsibilities, Records, and Reports

 TIP Responsibilities need to be in agreement with the provisions of GCP (the roles of the sponsor, the monitor, and the investigator).

- adherence to the protocol;

- complete, accurate, and current record keeping;

- adherence to the patient informed consent procedure;

- agreement on protocol amendment procedure; and

- commitment to accept monitoring visit by a designee of the company and to give access to source documents.

Article 4: Property/Publication

- agreement on publication strategy;

- nomination of a steering committee to review and approve manuscripts; and

- if patentable subject matter is present in the proposed publication (intellectual property protection SPON-SOR), the SPONSOR may delay any publication for up to 90 days after steering committee approval.

TIP Whereas it would not be ethical to oppose a publication, the manufacturer can include a clause stating that the manuscript will have to be reviewed prior to publication and that sufficient time is to be allowed for comments.

Article 5: Ownership of Confidential Information

- Statement on who is the owner and proprietor of all information, and that other parties have no right, title, or interest therein.

- Statement that information shall be kept confidential.

- Statement acknowledging that obligation of confidentiality does not include information already in possession, information in the public domain, information received from a third party, or information provided by law or by a regulatory agency.

Article 6: Indemnification and Reimbursement

- SPONSOR indemnifies INVESTIGATOR and INSTITUTION against liabilities imposed or authorised by law for adverse experiences to patients caused directly from the use of the DEVICE in the STUDY.

- Statement concerning terms for reimbursement of damages (e.g., reasonable costs of medical treatment to the extent that such costs are not covered by the patient's medical or hospital insurance or by third-party or governmental programs providing such coverage).

- Statement concerning there shall be no reimbursement for the adverse experiences of the procedure, nor for interventions that are deemed necessary in the clinical management of the patient.

- Statement concerning that SPONSOR indemnification/ reimbursement policy does not cover liabilities resulting from any misconduct, negligent act, or omission on the part of any indemnified party, or from the consequences of any inaccurate diagnosis made by any indemnified party relating to any subject's eligibility to participate in the study.

- Other conditions.

TIP │ Check local regulations on the legality of payment to investigators and institutions (e.g., French DMOS law, January 1993).

Article 7: Cancellation

- Conditions for cancellation from the SPONSOR.

- Conditions for cancellation from the INVESTIGATOR.

Article 8: Duration

- Date of the start of the agreement.

Article 9: Assignment

- no party shall have the right to assign, delegate, or transfer at any time to any party, in whole or in part, any or all rights, duties, and interests herein granted without first obtaining the written consent of the other parties.

Article 10: Severability

- if for any reason a court of competent jurisdiction finds any provision of the agreement, or portion thereof, to be unenforceable, that provision of the agreement shall be enforced to the maximum extent permissible so as to comply with the intent of the parties, and the remainder of this agreement shall continue in full force and effect.

Article 11: No Waiver

- failure by any party to enforce any provision of the agreement shall not be deemed a waiver of future enforcement of that or any other provision.

Article 12: No Warranties

- all warranties will be excluded by the SPONSOR, whether expressed or implied, by operation of law or otherwise.

 TIP | In order to avoid any potential financial dispute if the contract is not carried out until its end, define the status of the investigator as a subcontractor and not an employee of the company.

Article 13: Relationships of Parties

- parties are independent contractors.

Article 14: Dispute Resolution

- parties will try to resolve any differences on a friendly basis.

- parties will designate a representative who will meet with the parties and try to find a solution in case of dispute.

Article 15: Applicable Law

- this agreement shall be governed by and construed in accordance with the laws of . . .

Article 16: Notices

- all notices required or permitted under this agreement shall be in writing, with reference to this agreement, and be deemed given when . . .

Article 17: Complete Agreement

- the agreement consists of . . . , and supersedes earlier versions. No amendment to or modification of the Agreement will be binding unless in writing and signed by a duly authorised representative of all parties.

Signature Page

- names and addresses of involved parties.

Appendices

- monitoring procedure.

- unanticipated serious adverse events procedure.

WHEN TO DISCUSS THE CONTRACT DURING THE PREPARATION PHASE

The contract should be discussed once the investigator has decided to participate in the clinical study and has been selected and qualified by the monitor or sponsor. The investigator needs to know in detail what is expected from him/her. This entails that the investigator needs to know all the consequences with respect to the time required from investigator and staff, the facilities that will be needed, the administrative requirements, and the requirements to ensure GCP guidelines are adhered to.

Ideally, the monitor should send the draft contract to the investigator **two weeks** before the initiation meeting is organised. During this initiation or start-up meeting, both the investigator and the staff will be fully instructed and trained for the trial, and the contract with the principal investigator could be discussed; if no questions arise, then the contract could be signed by the investigator.

12

Informed Consent

Philippe Auclair
Guidant S.A.
Zaventem, Belgium

Every clinical investigation needs to comply with the provisions of the Declaration of Helsinki. Specifically, prior to being enrolled in a clinical investigation, subjects are required to receive easy-to-understand information about the clinical program from the investigator. This information should be comprehensive enough to enable the subject to understand the objective of the investigation, the expected benefits, as well as the risks and inconveniences associated with his/her participation.

The patient should have the opportunity to question the physician and have sufficient time to decide about his/her participating in the study, without coercion or duress. The decision to participate is documented by the subject's signature on the informed "consent form". The subject at all times has the opportunity to withdraw consent at any time without any sanction or loss of benefits to which he/she is otherwise entitled. Under normal circumstances, a subject cannot be enrolled if the informed consent has not been obtained.

HISTORICAL BACKGROUND

The moral, regulatory, and legal duty of obtaining an informed consent from each subject enrolled in a clinical investigation arises from the concept of *autonomy*. "Autonomy" is a term derived from the Greek *autos* ("self") and *nomos* ("rule"). In society, personhood is described in terms emphasising the individual's right, self-determination, and privacy; respecting the individual as a human being, in place of abusing or using it as a *means*.

In the past, the physician or surgeon was thought of in terms of a supposed expert in knowledge and skills that could not be questioned by the patient. This concept changed with the realisation that a human being could be included in a clinical investigation as a subject whether the person is being treated as a patient or is a healthy volunteer. This became a point of focus in the aftermath of the abuse of medical power by totalitarian regimes, such as the Nazis, or the abuses on prisoners in the United States in the 1950s. Regulations were drafted, beginning with the Nuremberg Code (1947) and followed by the Declaration of Helsinki (1964, and subsequent revisions).

The Nuremberg Code, for instance, states that in medical experiments on human beings, 'The voluntary consent of the human subject is absolutely essential'. No exemption is mentioned.

The Code further requires that participation in biomedical research is to be based on the freedom of choice, with no element of coercion or constraint. It also dictates that a subject should sufficiently understand the matter of the research to make an enlightened decision.

The Declaration of Helsinki, on the other hand, also requires informed consent. Should researchers believe that, in a specific case, an informed consent is impossible to obtain, they must give specific reasons concerning this to an independent committee. The Declaration of Helsinki also states that 'concern for the interests of the subject must always prevail over the interests of science and society.' Health care professionals are obliged by various ethical codes and laws to respect the intrinsic value of potential subjects as expressed by their 'autonomy'.

REGULATORY ASPECT

> Clinical investigations must be carried out in
> accordance with all laws to protect subjects
> (i.e., the Helsinki Declaration[1] or conventions
> on human rights in biomedicine). It is manda-
> tory that all measures relating to the protection
> of human subjects are carried out in the spirit
> of the Helsinki Declaration.

Regulations usually ask for compliance with the Declaration of
Helsinki or equivalent bioethics laws as a basic rule for obtaining the informed consent and ensuring rights of subjects are
protected.

In some countries some of the regulations may go a step further, defining additional requirements. For instance, the transposition of the European Directives for medical devices
introduce new requirements into national laws. Local laws are
not superseded by the directives; such as the following:

- FRANCE: Law 'Huriet' 88-1138 and subsequent decrees
  about protection given to subjects enrolled in clinical
  investigations

- ITALY: Decreto Ministeriale 27 April 1992 and Law
  Comma L 519/1973

- SPAIN: Decreto Real 561/1993

Non-compliance with Good Clinical Practice (GCP) could give
rise to huge penalties for sponsors and investigators as well as
potential litigation exposure.

LEGAL ASPECT

As previously stated, non-compliance with bioethics laws (see
chapter 1) can lead to potential legal action initiated by either

---

1. 18th World Medical Assembly in Helsinki, Finland, in 1964, amended by
the 48th General Assembly in South Africa in 1996.

the authorities or subjects seeking compensation. Before 1948, a claim for damages required evidence of negligence. In the specific medical context, a proof of negligence was impossible because of the alleged skills of the medical profession. This situation, which was clearly unfair to the patient, changed after the publication of the Declaration of Helsinki. After 1948, a claim for damages still needed to include evidence of negligence, but the proof became easier, premised on the legal principle of 'the thing speaks for itself'. This new situation, however, became unfair for sponsors or professional defendants. In a next step, negligence could be demonstrated by what could be paraphrased as 'being out of the state of the art as defined by similar professionals'. However, lawyers were not satisfied with attempts to recover damages without having to prove negligence. Since consent requires information on which to base a decision, lawyers suggested that only an 'informed consent' would be an adequate defence against being sued. Adequacy of disclosure of the information on which the consent is based is now considered a key factor.

It was commonly understood that consent forms cannot be contracts. This assumption may, however, prove to be incorrect based on some recent rulings: An informed consent form can create a contract between the sponsor and the subjects. The resulting implications of these rulings is that sponsors in drafting informed consent forms must not only consider the ethical issues, but also take care not to inadvertently create a contract. Any non-fulfilled promise stated in an informed consent can be an opportunity for legal action (e.g., the manufacturer agrees to provide the device to study patients after the study has ended, and cannot deliver since pre-market approval is denied).

LOCAL ASPECTS OF CONSENT

Attempts so far to harmonise the ethical aspects of clinical investigations has failed. Practical implications are that multi-centre and multi-national clinical investigations cannot rely on a single Ethics Committee or Institutional Review Board (IRB) approval, nor on a single informed consent form. National regulations as well as geographic and cultural differences form the obstacles.

Language is the most obvious one. How can a consent be 'informed' if the subject signs a form that is not written in his/her native language? This could even be considered a violation of the basic principles of bioethics (e.g., the Declaration of Helsinki). It is evident that the consent form needs to be written in the local language of the country where the clinical site is located.

The situation, however, is further complicated by the high mix of ethnic populations living in Western countries. Not all of the subjects being treated in a given hospital may be fluent in the official national language. Migrant workers and reimbursement issues further complicate the matter. This can lead to difficulties in ensuring that the consent has been properly understood, not to mention the legal consequences which could ensue if a subject seeking compensation maintains, in a court of law, that the consent form was not understandable.

Enforcing the principle of autonomy and the requirements of informed consent presents some difficulties in countries where personal choice is restricted. For example, in some African countries, the concept of personhood has a totally different meaning than in countries of western Europe. Individualism is not defined *per se* and the person can only exist through collectivity. In some cases, the issue of who is responsible for giving the consent (i.e., the head of household, elders, individual persons, the chief of the community, or even the government) needs to be discussed before a study is put in place. Cultural sensitivity is required.

This contradicts the opinion that a study is ethical or it is not. Why should it be believed that the ethical principles of one culture are valid in another? The result is that a multi-centre investigation can end up with as many versions of informed consent as there are centres involved in the investigation, after review and modification by each hospital or regional Ethics Committee or IRB.

MINIMUM INFORMATION REQUIRED

Basic Elements of Informed Consent

In seeking consent, the following information should be provided to each subject to conform with the provisions in the FDA *Federal Register* and various European laws and guidelines:

- a statement of what the study involves, an explanation of the purposes of this research, and the expected duration of the subject's participation together with a description of the procedures that are related to the experiment (including the number of visits, special procedures, etc.);

- a description of any reasonably foreseeable risk or discomfort to or for the subject;

- a description of any benefit to the subject or to others what may be expected;

- a disclosure of the appropriate alternative procedures or treatments, if any, that may be advantageous to the subject;

- a statement describing the extent, if any, to which the confidentiality of records identifying the subject will be maintained and denoting the possibility that the administration may inspect the records;

- if the study involves more than minimal risk, an explanation as to whether any medical treatment is available if injury occurs and, if so, what is involved or where further information may be obtained;

- an explanation of whom to contact for answers to pertinent questions about the research and subjects' rights, and whom to contact in the event of a research-related injury to the subject;

- a statement that participation is voluntary, that refusal to participate will involve no penalty or loss of benefits to which the subject is otherwise entitled, and that the subject may discontinue participation at any time without penalty or loss of benefits to which he/she is otherwise entitled;

- a clause by which the subject will give permission for the inspection of hospital files by the regulatory authorities and authorised company personnel; and

- a statement that an appropriate insurance policy has been arranged for the subjects.

Additional Elements of Informed Consent

In addition to the basic elements, the following elements of information should also be included in or supplement the informed consent:

- a statement that the particular treatment or procedure may involve risks to the subject (or to the embryo or foetus, if the subject is or should become pregnant) that are currently unforeseeable;

- anticipated circumstances under which the subject's participation may be terminated by the investigator without regard to the subject's consent;

- additional costs to the subject that may result from participation in the research;

- the consequences of the subject's decision to withdraw from the research and procedures for orderly termination of participation by the subject;

- a statement that significant new findings developed during the course of the research that may relate to the subject's willingness to continue participation will be provided to the subject; and

- the approximate number of subjects involved in the study.

GUIDELINES FOR THE INFORMED CONSENT OF TRIAL SUBJECTS[2]

- In obtaining and documenting informed consent, the investigator should comply with the applicable regulatory requirement(s) and should adhere to GCP and to the ethical principles in the Declaration of Helsinki. Prior to the beginning of the trial, the investigator should have the IRB or Ethics Committee's written approval/favourable opinion on the informed consent and written information to be provided to the enrollees.

- The informed consent and written information to be provided to the subjects should be revised whenever new information becomes available which may be relevant to the subject. Any revised informed consent and written information should receive the IRB's or Ethics Committee's approval in advance of use.

- Neither the investigator nor the trial staff should coerce or unduly influence the subject to participate or to continue to participate in a trial.

- None of the oral and written information, including the informed consent, should contain any language that causes the subject or subject's legal representative to waive or to appear to waive any legal rights, or that releases or appears to release the investigator, the institution, the sponsor, or their agents from liability for negligence.

- The investigator, or person designated by the investigator, should fully inform the subject. Should the subject be unable to provide informed consent, such as in cases of minors or comatose patients, the subject's legal representative needs to be informed of all the pertinent

---

2. Adapted from the International Conference on Harmonisation of Technical Requirements for the Registration of Pharmaceuticals for Human Use. ICH EWG E6: Good Clinical Practice—Draft 19; 27 April 1995, pp. 21–25.

aspects of the trial, including the written information as approved by the IRB/Ethics Committee.

- The language used in the oral and written information, including the informed consent, should be non-technical and should be understandable by the subject or the subject's legal representative.

- Before informed consent is given, the investigator, or a person designated by the investigator, should provide the subject or the subject's legal representative sufficient time to decide whether or not to participate in the trial and the opportunity to inquire about trial details. All questions about the trial should be answered to the satisfaction of the subject or the legal representative.

- Prior to participation in the trial, the informed consent form should be signed and dated by the subject, or the subject's legal representative, as well as by the person who conducted the informed consent discussion.

- If the subject or subject's legal representative is unable to read, an impartial witness should be present during the informed consent discusion. The informed consent form should be read aloud to the subject, whereby the subject needs to consent orally to having understood the contents. The subject or subject's legal representative and the impartial witness as well must each sign and date the form. The impartial witness is required as an attest that informed consent was freely given by the subject or subject's legal representative.

- Both the informed consent discussion and the informed consent form should include clear explanations

  - of the research involved in the trial;
  - of the purpose of the trial;
  - of the trial treatment(s);
  - of the trial procedures to be followed, including all invasive procedures;

- of the subject's responsibilities;

- of the trial features that are experimental;

- of the reasonably foreseeable risks or inconveniences to the subject;

- of the reasonably expected benefits. When there is no intended clinical benefit to the subject, the subject should be made aware of this;

- of the alternative procedure(s) or course(s) of treatment that may be available to the subject, and their potential benefits and risks;

- of the compensation and/or treatment available to the subject in the event of trial-related injury;

- of the anticipated pro-rated payment, if any, to the subject for participating in the trial;

- of the anticipated expenses, if any, to the subject for participating in the trial;

- that subject's participation in the trial is voluntary and that the subject may refuse to participate or withdraw from the trial, at any time, without penalty or loss of benefits to which the subject is otherwise entitled;

- that monitor(s), auditor(s), the IRB/Ethics Committee, and regulatory authority(ies) will be granted direct access to the subject's original medical records for verification of the clinical trial procedures and/or data without violating the confidentiality of the subject to the extent permitted by the applicable laws and regulations. By signing an informed consent, the subject or the subject's legal representative authorises such access;

- that the confidentiality of records that identify the subject will be maintained and will not be made public to the extent permitted by the applicable laws and/or regulations;

- that the subject or the subject's legal representative will be informed in a timely manner if information becomes available that may be relevant to the subject's participation in the trial;

- of the person(s) to contact for further information regarding the trial, the rights of trial subjects, and whom to contact in the event of a trial-related injury;

- of the circumstances under which the subject's participation in the trial may be terminated without the subject's consent;

- of the expected duration of the subject's participation in the trial; and

- of the approximate number of subjects involved in the trial.

- The subject or the subject's legal representative should receive a copy of the signed and dated informed consent as well as all updates of the consent form, and a copy of the written trial information and relevant amendment(s).

- To the extent that a minor child is capable of understanding and granting informed consent, the minor should give consent as well as the minor's legal guardian. Both the minor, if so able and capable, and the minor's legal guardian should sign and personally date the informed consent.

- In a non-therapeutic trial (i.e., when there is no anticipated direct clinical benefit to the subject), consent must always be given by the subject and the informed consent must be signed and dated by the subject.

- Where prior consent of the subject is not possible, the protocol is submitted to the IRB/Ethics Committee. The IRB or Ethics Committee may provide that such consent need not be obtained and that only the consent of the subject's legal representative, if available, should be solicited. In case of absence of the subject's legal representative, other measures are to be taken which need to be defined in the protocol.

THE PROCESS OF OBTAINING INFORMED CONSENT

Most medical ethicists view informed consent as a process in which the investigator communicates to the subject the significance of their participation in a clinical trial.[3] Giving consent is not merely an act of signing a document. It is a process through which the subject becomes informed about the clinical investigation, and clearly understands that he/she is participating in a research protocol. Although informed consent ideally involves both verbal and written statements, the evidence that consent has been obtained is the signature of the subject on the designated form.

The nature of the information given to the subject can be extremely complicated. Verbal communication becomes extremely important. The information provided to the subject should explain

- the potential risks and benefits expected from participating in the clinical investigation;

- the length and nature of the clinical investigation, and number of subjects in the clinical investigation;

- the patient's rights, including that participation is voluntary and that withdrawal can be done at any time;

- who the subject should contact if any questions about the research or their rights as research subjects arise;

- the patient's responsibilities as a research subject, including responsibilities for follow-up visits for data collection;

- that the patient has enough time to consider whether he/she wants to participate in the trial (e.g., not having a consent signed the day before surgery); and

---

3. *Investigational Devices: Four case studies.* Department of Health and Human Services, April 1995.

- as necessary, that the device has not been approved as safe and effective.

THE READING DIFFICULTY OF INFORMED CONSENT DOCUMENTS

Informed consent documentation can range from a short and non-informative document to a long, convoluted, and difficult-to-understand document. Vague information can obviate a consent procedure. On the other hand, documents which are long and complex will not be read by any subject.

The verbiage involved, including technical names for medical devices and related procedures, contributes to a subject's confusion. An absolute necessity is clear information that is understandable by the layman. A perfect informed consent does not exist. The question remains, therefore, whether patients understand what they sign, and if not, is consent then of any use?

SPECIFIC CASES

In specific cases (i.e., emergency, children, incapacitated person), consent cannot be given before inclusion in the study. Several options have been suggested as 'differed consent', whereby the consent is collected retrospectively when the subject is able to understand. It is clear, though, that in such situations the subject cannot refuse to participate. A consent by proxy can be used whereby a guardian or legal representative in such situations signs on behalf of the incapacitated subject.

Pragmatically, a proxy consent is deemed acceptable when[4]

- there is no other reliable or relevant route to the same end;

- the end in some sense justifies the risk; and

---

4. Dunstan, G.R. 1987. Evolution and Mutation in Medical Ethics. In *Ethical Dilemma in Ethical Promotion*, ed. S. Doxiadis. Chichester, UK: John Wiley & Sons.

- the subjects are not treated merely as things.

Absence of consent may be justified when

- a life-threatening situation exists that necessitates the use of the device;

- there is no effective alternative to the use of the device;

- it is not feasible to obtain the informed consent from the patient; and

- there is insufficient time to obtain informed consent from the patient's legal representative.

In the latter case, the investigator must obtain the written confirmation of the aforementioned by a licensed physician not involved in the investigation. It is highly recommended for the sponsor exposed to these problems to include these provisions in the investigator's agreement in order to prevent any further dispute that may arise after a patient is included without giving proper consent.

CONCLUSION

Some concerns, in a few cases, have been raised that some human subjects enrolled in clinical research may have been exposed to unacceptable risks. As a consequence, the World Medical Assembly issued the guideline known as the Declaration of Helsinki. It is a basic requirement of GCP that clinical investigations be carried out in the spirit of the Declaration of Helsinki. The principles expounded herein have found their way into legal instruments, moving the obligation to protect rights of subjects and inform prior to enrollment into the realm of bio-law, with enforceable sanctions in cases of non-compliance.

Before taking part in a clinical investigation, the potential subject is to be informed about the purpose of the investigation, and the associated procedures, as well as any discomfort or inconvenience that may be involved. A genuine consent is to be obtained without pressure or duress. The resulting form, signed by the patient, may vary according to national regulations and uses. It almost always needs to be in a language and verbiage understandable to the subject. This consent has to be obtained prior to the subject's inclusion in the study. A range of issues— moral, historical, legal, and professional—are to be taken into account when designing an informed consent statement.

13

The Design of Case Report Forms

Herman Pieterse

Profess® Medical Consultancy BV
Heerhugowaard, The Netherlands

Peter Duijst

Aortech Europe Ltd.
Almere, The Netherlands

In a clinical study, Case Report Forms (CRFs) can be defined as the set of documents that have been specially designed for use in a study and that aim for the complete recording of all data relevant to the patient and/or the treatment applied. The CRFs, which represent an essential part of the study protocol, should be strictly in line with the study design. In other words, they need to be a true reflection of the objectives of the protocol.

Thus, it is imperative that CRF design matches the protocol to such an extent that ultimately the right data will be collected to allow a valid analysis. Incorrect design of the CRFs will lead to the collection of irrelevant data and obscure the essentials of the trial.

In general, the function of CRFs is to provide the exact answer to questions posed by the study protocol and to provide the required safety data regarding the product being studied.

| NOTE | The close harmony between the study protocol and the CRF makes the parallel development of the two the preferred option, rather than a sequential development scheme. Simultaneous generation will facilitate a process of creating congruent designs for both the protocol and the CRFs. |

The CRFs will also need to link up with the flow of events in the clinic. Such a design, following clinical reality as much as possible, will generate logical and transparent forms for the investigator. By avoiding confusion as much as possible, this approach will definitely have a positive effect on the quality of collected data.

Whenever possible, standardisation in design for the various CRFs should be stimulated, without avoiding prevalance over an individual approach per study. A delicate balance should be kept between standardisation and the individual needs for each trial.

WHO IS INVOLVED IN THE DESIGN?

Many people representing different disciplines need to be involved in the CRF design. Only in this way can a balanced representation of the various demands put on the design (i.e., the CRFs need to be practical in use, uncomplicated in their layout, and allow for the collection of the pivotal data) be expected and achieved.

The direct implication of the aforementioned is that a TEAM will have to be formed to deal with the different aspects of the design. For every team, management will be necessary to orchestrate the activities. Having a number of different disciplines represented in the team will make the management a complex job, which demands a skilled manager (see chapters 3 and 6). Usually, the clinical project manager or a study monitor will be appointed for this. The CRF design group will consist, at a minimum, of the persons listed in Table 13.1.

The consequence of implementing a design procedure through a team, of course, will be that time needs to be invested by all of the participants. However, the time spent in this phase of the clinical trial will pay off later. The overall efficiency of the trial will gain considerably from well-designed CRFs. The

Table 13.1. Overview of the Members and Their Respective Input in the CRF Design Group

CRF Design Team Members	Expected Input
Clinical project manager	Manages the CRF design project and controls the communication between the different parties involved.
Investigator	Sees to it that the forms match the flow of clinical events.
Monitor	Sees to it that the forms match the study protocol requirements.
Data entry clerk	Makes sure that the final CRF design will be practical for the facilitation of correct data entry.
Statistician	Oversees the process of data analysis from the CRFs.
Form designer	Controls the layout aspects of the forms.
Database software expert	Sees to it that the CRF design can be translated into appropriate software. Not uncommonly, this is also part of the responsibilities of the statistician.
Statistical software expert	Deals with adaptations in the statistical software when it is necessary to accommodate the CRF design. Not uncommonly, this is also part of the responsibilities of the statistician.

fact that everybody involved has been part of the design process will increase the acceptance of the forms and reduce the level of erroneous use.

THE DEVELOPMENT PROCESS

The first important step to take will be to nominate the CRF design group. This should be done by the person who has been appointed the clinical project manager once senior management has approved the clinical trial. The initiation meeting of the group will deal mainly with the strategy and logistics that are described below.

Since in many clinical trials a basic set of the same data is collected, it makes sense to standardise the CRFs as much as possible. If this has not been company policy until now, this approach should be implemented at this point.

Not so much for CRFs under design, but in particular for follow-up studies, it will be convenient to have a set of standard CRFs in place, forms with which the investigator and especially the monitor are already familiar. The more often standardised CRFs can be used in different studies, the more efficient their actual design becomes. Either the CRF design group should start making use of existing CRFs or start developing forms that allow for standardisation. The CRF design group should also determine the procedure used to change standardised forms.

Once the design group has been installed and the first meeting held, the pace for further development can be set. The key issue will be that all members are adequately informed on the status of the design programme and the progress being made. This means that emphasis will have to be put on communication within the group. Typically, this is the responsibility of the project manager.

| NOTE | The worst situation that may occur is one where everybody is talking to everybody about every single issue without others knowing about it. This form of communication will not contribute to an adequate exchange of information in a group, but lead to a situation where nobody is really being informed! |

Outside group meetings, when information will need to be exchanged between group members, information should always come to and be distributed by the project manager to the other group members. In this way, all members will be equally informed about the project.

NOTE Communication also means that senior management is adequately informed on a regular basis about the status of a project. Not only should progress be reported, but also problems and delay. It is in nobody's interest to report negative issues in a late phase when corrective action is much more difficult to implement!

Apart from the communicative issue discussed previously, another important issue to address is maintaining the initial enthusiasm of the group members for the project. Once the project has successfully started, the momentum should be kept going.

Step 1: The project manager consults all parties involved.

Step 2: A check is done on the availability of existing, applicable, standard CRFs.

Step 3: The first CRF draft is prepared and sent for review to all parties involved.

Step 4: The first draft and comments are discussed at the design group meeting.

Step 5: The minutes of the meeting are given to all parties involved, including senior management.

Step 6: Repeat steps 3 to 5 for succesive CRF versions.

Step 7: Repeat steps 3 to 5 for final CRF version.

Step 8: Design manager officially informs the design group members and senior management of the outcome.

One of the best ways of arranging this is to make sure that the frequency of meetings is planned in advance and known to everyone in the group, and that minutes of project meetings will be generated and provided speedily to the members.

DESIGN ISSUES

For most clinical trials, the CRF design will focus on a number of crucial issues. The data flow needs to closely follow the clinical reality in order to accommodate the investigator as much as possible. This will increase the lucidity of the forms for the investigator and ease the use of them.

TIP	The most important issue for the design is to ensure the highest possible quality of initial data capture. The interest of the persons processing and analysing the data must be secondary—but will be served well by the collection of high-quality data.

When following the clinical flow of events in the CRF design, it should be realised that three different forms of clinical data are usually collected.

1. *Efficacy data,* which answer the question, "Does the device perform according to the specifications given?" In many cases, these data are device specific.

2. *Safety data,* which picture the safety profile of the device for the individual subject, and usually comprises adverse events and laboratory data.

3. *Background data,* which include various details relating to the subject's demographics, physical status, and concomitant medication during the course of the study. These background data are essential to putting the efficacy and safety data into context.

Whenever possible, the CRFs should be designed so that the respective forms represent only one class of data, that is, only efficacy, safety, or background data. This will create coherent groups of data (CRF modules), which are related to each other. For this reason, they are comprehensible to the user and facilitate data entry and analysis in a later stage. Of course, whenever the design team foresees that this design approach may compromise the regular flow of clinical events for the study subjects, it will have to be adjusted or even entirely abandoned.

Another important advantage of designing CRF modules that represent a class of data rather than a mixture of data is that the forms can be easily compiled into sets of forms. The investigator will have one set of CRFs for each of the subject's visits.

 Laboratory forms at the end of the CRF package may lead to confusion with the investigator on whether the data are requested for a particular visit, or the investigator may overlook the forms entirely and forget to provide the data.

The forms should also avoid being confounding to the user. The forms should lead the investigator rather than forcing the investigator to look for where to enter the data. Above all, the CRFs need to be easy to understand. A number of important basic rules in CRF design are given below:

- *Limit the amount of data to be entered on the CRFs to those that will answer the questions formulated in the study protocol.* When nonrelevant data are requested, the investigator will become irritated and lose his/her interest in the study.

- *Limit the amount of data per form.* The more data that are requested on the CRF, the more confusing the form becomes, and the more time it will take to complete. A well-designed form can be easily and quickly filled out. Having more forms to complete is usually no problem for an investigator, as long as the pace can be kept.

- *Limit open text responses or space for comments to a minimum.* For purposes of data entry, comments and other text need to be converted into a code. The more text given, the more complicated this becomes.

- When open text responses must be included, use a line with character spaces, for example,

It will definitely encourage the investigator to write more legibly and concisely, but, on the other hand, may cause irritation and confusion as well.

- *Do not repeat the same questions for successive visits of the study subjects, unless the answers may differ.* It does not make sense to ask for information that is already known.

- *Try to organize preprinted CRFs with basic repetitive information, such as the investigator's name, clinical site, visit code, and visit numbers.* For the investigator, preprinted forms will be quite convenient and time-saving.

- *Multiple choice questions, such as yes/no or none/ mild/moderate/severe, should be in the same order throughout all CRFs.* Changing the order of options in questions easily leads to mistakes by the investigator.

- *Standardise the CRF forms as much as possible.* This is an important tool for controlling data collection in successive studies with the same investigator(s). The individual(s) will already be familiar with the forms. For the process of data entry and analysis, standardisation will be a major help.

- *Use ticking boxes to answer questions.* This makes the forms orderly in arrangement and easy to handle.

- *Avoid placing the ticking boxes before the questions.* The natural order in communication is that the question precedes the answer.

- *Use boxes for numerical data (laboratory, etc.).* In this way, a standard format can be imposed on the number of decimals.

- *Select a font and type size carefully.* A well-chosen (i.e., distinct) font and size (as long as it's not small) make the text easy to access and the forms pleasant to work with.

- *Allow an adequate margin for binding.* Especially when the CRFs are presented in a binder for each study visit, it may be difficult to detect what has been printed in the margin.

- *Use shading sparsely.* Shading may obscure parts of the CRF, and it does not photocopy or fax well.

- *Be careful when asking the investigator to fill in codes in the checkboxes.* Mistakes are easily made here, particularly when the investigator is too occupied to realise that more attention should be given to selecting the correct codes.

- *All forms must have a clear space for the investigator's signature and for completing the date of signing.* Reserving space for the co-investigator's signature also makes sense when this is the person who will be completing the forms.

CRFs designed by only the study monitor or the project manager tend to lead to the generation of impractical forms, lacking essential input from others who will also work with the forms. Knowing how others will use the CRF makes designing them a less complicated matter. All parties for whom the forms are intended should be involved in the design as much as possible. Again, the information exchange will play a crucial role in determining the final result.

It is important to realise, however, that whatever thoughts may have been expressed on this topic or whatever claims have been made, the ideal CRF has not yet been designed and probably never will be designed!

WHAT SHOULD A SET OF CRFS FOR AN INDIVIDUAL STUDY LOOK LIKE?

For most clinical trials, the set of CRFs applied will be basically congruous. Considering that the CRF contents should be in line with the study protocol, a document that is standardised to a large degree, this is not very remarkable.

A distinction can be made between two CRF types: generic CRF modules, which are general in design and form the basic CRFs in a trial, and study-specific CRFs, which have been specially designed for a particular trial. In contrast to the study-specific CRFs, the generic ones may demand only slight adaptation or none at all for use in other trials because the fundamental design pattern always remains the same. For the study-specific CRFs, this will take more effort but still will be worthwhile in many cases. Table 13.2 outlines the types of generic and study-specific CRFs.

The advantage of designing CRFs in a modular fashion is that for each visit, sets of the applicable forms for that specific visit can be compiled and presented to the study site in binders. This will make the whole process of providing CRFs much more orderly and potentially less obtrusive to the investigator, who will not have to worry whether all of the necessary forms have been filled in at the end of the subject's visit.

STANDARDISATION

This chapter has frequently mentioned the issue of CRF standardisation, while only briefly going into the background of it. It is not difficult to see that this method of CRF design holds a number of obvious advantages for all parties involved in clinical research. On the other hand, it should also be understood that the standardisation of CRF design will only gradually result in more practical and accessible forms, with an increased quality of the data collected.

Standardisation is a continuing process that runs over a sequence of studies, with each of the designs benefitting from the experience previously obtained in earlier studies. Implementation will preferably be done by a team specifically chosen for this purpose and having strong overlap with the various CRF design groups. Ideally, the intention will be to develop CRFs that can be used in a multitude of clinical trials without the need for redesign, or requiring only minor adaptations. For the investigator, the monitor, the data entry clerk, and the

Table 13.2. Types of Generic and Study-Specific CRFs

Generic CRF Modules	Study-Specific CRFs
MEDICAL HISTORY	ELIGIBILITY
Pre-study medical evaluation of the subject outlines the pathology profile.	Outlines the protocol inclusion and exclusion criteria. Should end with the conclusion that the subject is yes or no for the study.
VITAL SIGNS	MEDICATION
Basic medical parameter values to be obtained during subject visit (e.g., blood pressure, weight, body temperature, etc.).	Drugs that have been prescribed specifically for the clinical condition(s) being monitored in the device study.
PHYSICAL EXAMINATION	SPECIFIC LABORATORY TESTS
Findings from the routine physical inspection of the various tracts (e.g., respiratory, circulatory, gastrointestinal, etc.).	The collection of objective data from clinical investigations strictly related to the device (e.g., ECG, EEG, ECHO, etc.).
CONCOMITANT MEDICATION	SPECIFIC SCALES, QUESTIONNAIRES
Drugs prescribed for clinical conditions other than those monitored in the device study, and which may interfere with the study (e.g., diabetes may influence the outcome in an ophthalmological trial).	Subjective data collected to express the subject's clinical condition according to a clinical classification system (e.g., NYHA [New York Heart Association] classification in cardiac failure).
INFORMED CONSENT	DEVICE PROCEDURE
Records the date of the subject's approval and reminds the investigator.	Specifically deals with the data to be collected during the device's clinical application.
SUBJECT WITHDRAWAL	ADVERSE EVENTS
Presents the date and reason for the subject's withdrawal. Should be part of each visit's set of forms.	Form designed to record all complications during the clinical trial, but in particular focussing on the device and/or procedure-related incidents.

statistician, the major beneficial issues linked to working with standard CRFs are detailed in the following sections.

| NOTE | Primarily, the standardisation process should lead to CRF modules that follow the clinical flow of events rather than being a convenience to the persons entering and analysing the data. Their interest will be served best by the collection of high-quality data.

Advantages to the Investigator

Especially when the investigator has been involved in previous studies with previous generations of the forms, he/she will have grown accustomed to the layout of the forms and will have little difficulty working with the CRFs. When the standardisation process results in well-designed forms with an easy-to-access layout, the investigator will find that the forms are convenient and easy to complete, even when using them for the first time.

Advantages to the Monitor

The arguments for the investigator also apply here. The result will be that the investigator will require less explanation; because less incorrectly entered data may be expected, the monitor will have to take less corrective action. The number of monitoring visits may be reduced, and the efficiency of the monitoring process will increase.

Advantages to the Data Entry Clerk

Having standardised CRF modules in the database will lead to a standard data entry procedure involving standard methodologies of coding data for the studies in which the CRF modules have been applied. Since the data entry procedure will have become basically uniform for the various studies performed, this will lead to more accurate data entry and, ultimately, generate higher quality databases in a shorter period of time.

Advantages to the Statistician

Once CRF modules have been developed, databases can be set up more rapidly and more reliably. Standard programs to check for missing, inconsistent, and dubious data can be used repeatedly without the need for an extra validation of the program. Another advantage will be that listing, tabulating, and analysing the study data can be done with more ease and speed. Comparison across studies for the purpose of providing, for example, safety data updates to satisfy regulatory requirements can be done more quickly and consistently as well.

REFERENCES

'Good Clinical Practice for Trials on Medicinal Products in the European Community'. 1990. *Pharmacology and Toxicology* 67:361–372.

Meinert, C. L. 1986. *Clinical Trials: Design, Conduct, and Analysis.* London: Oxford University Press.

Pocock, S. J. 1983. *Clinical Trials: A Practical Approach.* Chichester, UK: John Wiley and Sons.

The Scrip Guide to Good Clinical Practice: Documenting Trials to U.S. FDA Requirements. Richmond, Surrey, UK: PJB Publications Ltd.

14

Labelling and Instructions for Use

Robin N. Stephens

AVE UK Ltd
Burgess Hill, West Sussex, UK

Marja G. de Jong

Medidas
Wijk bij Duurstede, The Netherlands

1: European Labelling Requirements

THE AIM OF CLINICAL INVESTIGATIONS

The European Medical Device Directive (MDD) (93/42/EEC) in article 2.1 Annex 10 defines the aims of a clinical investigation with medical devices.

The objectives of clinical investigations are:

- to verify that, under normal conditions of use, the performance of the devices conform to those referred to in Section 3 of Annex I (the Essential Requirements [ERs]); and

- to determine any undesirable side effects, under normal conditions of use, and assess whether they constitute risks when weighed against the intended performance of the devices.

For the purposes of this chapter, the key phrases from this passage are as follows:

- "under normal conditions of use";

- "conform to . . . Section 3 of Annex I"; and

- "intended performance of the devices".

These phrases are important as they direct the medical device manufacturer to the need for clinical investigations to be performed with devices that, in as near a way as possible, reflect the actual products that are to eventually be placed on the market. The devices must be used, as far as is practical, under "normal conditions of use" so that accurate assessment of the observations, data, and results of the clinical investigation may be ensured. Also, it enables the extrapolation of these results to the large-scale production and marketing phase with as few errors as possible. Further discussion follows under "Conditions of Clinical Investigations" in this chapter.

The second and third key phrases are two parts of the same subject. As previously noted, the objectives for clinical investigations include "the performance of the device must conform to those referred to in Section 3 of Annex I". Annex I of the MDD has the general heading "Essential Requirements". Within this broad topic are two subsets: (1) "General Requirements" and (2) "Requirements Regarding Design and Construction". Section 3 falls within the former subset and deals with the performance of a device as specified by the manufacturer. Throughout the directive, be it related to device classification or to the Essential Requirements (ERs), the performance specifics given by a manufacturer is the determining factor. The intended purpose of a device is crucial to the correct understanding and interpretation of, and effective compliance with, the Directive. It is the manufacturer, the company, or the person placing a product on the European market under their own name, who *has the responsibility to determine the performance* criteria required for a device *to ensure* that it fulfils the intended purpose as stated by the manufacturer. The two are inextricably inter-related and need to be carefully considered before any claims—clinical or otherwise—are made for a medical device.

Section 3 of Annex I states:

> The devices must achieve the performances intended by the manufacturer and be designed, manufactured and packaged in such a way that they are suitable for one or more of the functions referred to in Article 1 (2)(e), as specified by the manufacturer.

As this chapter specifically deals with labelling and instructions for use that accompany medical devices, the most significant parts of this section are those relating to design and packaging. Design is significant because a device must be developed in such a way that it meets the user's needs and may be easily and effectively used. To achieve this, adequate instructions need to be incorporated with the product from the design phase onward, developing and evolving as the device takes shape toward its final format. Packaging is significant because the design characteristics of sterility, protection from damage, and so on need to be incorporated into the labelling that provides useful information on storage conditions and so on.

NOTIFICATION TO COMPETENT AUTHORITIES

Article 15 of the MDD requires manufacturers of devices intended for clinical investigation to "follow the procedure in Annex VIII and notify the competent authorities of the Member States in which the investigations are to be conducted". Annex VIII is entitled "Statement Concerning Devices for Special Purposes". It contains the requirements for a statement that must be drawn up by the manufacturer and presented to the Competent Authority prior to the commencement of a clinical investigation. These requirements include supplying:

- "data allowing identification of the devices in question and

- a statement that the device in question conforms to the Essential Requirements [i.e., Annex I] apart from the aspects covered by the investigations . . . "

The former is obviously the name of the device, any proposed trade name or trade-mark, an identification code or reference number, and serial number, if appropriate. This information must be supplied to the Competent Authority as part of the notification package, but must also be included on the device or on the labelling of the device so that identification is possible.

The latter statement refers to devices complying with the ERs in so far as the ERs are not the subject of the investigation. By definition, the requirements relating to labelling and instructions for use (Section 13 of Annex I) must be complied with regardlessly. See "Indirect Requirements of the Medical Device Directive" of this chapter for more details.

CONDITIONS OF CLINICAL INVESTIGATIONS

The objectives of clinical investigations include the need to test the product in the environment in which it will normally be used. The principle behind this is good science: In any experiment, one minimises the number of variables to enable the accurate observation of effects associated with the factors under study and the effective extrapolation to wider-ranging conclusions. In the case of medical devices, it needs to be demonstrated that the ERs are complied with; the device, when in normal use, is not to threaten a patient's health beyond an acceptable risk limit and provide the treatment, diagnosis, and so on that is intended. This is the reason for the references to "normal circumstances" that occur frequently throughout the Directive.

Obviously, under normal conditions of use, the medical device will be supplied in adequate packaging with labelling that (1) communicates useful information to the medical staff (i.e., product type, sterility details, size, etc.) and (2) conforms with the ERs as set forth in Section 13 of Annex I of the MDD. A suitable instruction leaflet containing precautions, contra-indications, technique advice, physical specifications, and so on should also be supplied. This, too, needs to be in conformity with Annex I.

This idea of instructions as per normal conditions is also found in GCP.

> The proposed objectives of the clinical investigation plan should be related to establishing or verifying the safety and performance of the devices, when used for its intended purpose and according to the documented instructions. (EN540 Article 5.3.4.)

Whilst the main thrust of this clause is the use of a product in conformity with the clinical investigation protocol, it includes intrinsic instructions for use accompanying the device under investigation, whether these instructions are part of the product package or the protocol itself. In fact, the instruction leaflet should be included with both the product's packaging and the protocol.

DIRECT REQUIREMENTS OF THE MEDICAL DEVICE DIRECTIVE

Section 13.3(h) of Annex I

Annex I of the MDD contains an article (13) that specifies the requirements for the labelling of medical devices (see "Labelling" in this chapter). Within these ERs Section 13.3(h) applies uniquely and specifically to devices intended for clinical investigations. Quite simply, it states that products for use in clinical investigations must be labelled *exclusively for clinical investigations.* This is to be placed clearly and prominently on the product's label.

Exemption from CE Marking

Only two groups of medical devices are allowed to move within the Member States of the European Union without CE marking: custom-made devices and devices for clinical investigations. This exemption from CE marking brings with it requirements for compliance with other relevant parts of the MDD (Article 15 and Annex X), but the overall thrust is that devices for clinical investigations do not require CE marking. Actually, it would be an offence to CE mark devices that, by definition of being intended for a clinical investigation, do not meet the ERs. Thus, whilst the purpose of clinical investigations is to provide data to verify safety of intended use and so on to enable a manufacturer to affix a CE mark, the devices utilised in these investigations do not require to bear a CE mark.

INDIRECT REQUIREMENTS OF THE MEDICAL DEVICE DIRECTIVE

Labelling (Section 13 of Annex I)

As has already been outlined, the MDD requires that all products intended for investigational use must meet all the ERs that are not under investigation in the studies to be performed. This means that, apart from requirements relating to design, manufacture, biocompatibility, and so on, the devices must also be labelled in compliance with Section 13 of Annex I. This section sets out the general information that must accompany a device, either on a label or as instructions for use.

The general principle of Section 13 is that "each device must be accompanied by the information needed to use it safely and to identify the manufacturer . . .". This encompasses both the information on the label and the instructions for use.

The specific requirements for the label on a medical device include

- the name and address of the manufacturer;

- the name and address of the authorised representative in Europe for products imported into the European Union;

- device identification (i.e., catalogue or item number, product name, description, etc.);

- information on the contents of the packaging (i.e., number of units, details of the contents of a kit, etc.);

- the word *sterile* (if appropriate);

- the lot number preceded by "LOT" (if appropriate);

- the serial number (if applicable);

- the date of "safe-use-before" (i.e., use by or expiry date);

- a "single-use" statement (if appropriate);

- the designation: "exclusively for clinical investigations";

- special storage or handling conditions, if relevant;

- appropriate operating instructions;

- warnings and precautions;

- the year of manufacture for an active device that does not have a use by date; and

- the method of sterilisation for a sterile device.

All of these basic characteristics must be included on any device label, as appropriate. Thus, a sterile cardiovascular catheter (for example) would at a minimum be labelled something like Figure 14.1.

Of course, devices utilised for clinical investigations will not normally be mass-produced items; the actual format and livery of the labelling employed may well not be the same as the ultimate finished product. However, labels in a format as near to the final intended version should be used, in so far as is possible, thus ensuring that in this respect at least the device is in its normal conditions for use. This may mean the inclusion of colour, logos, and symbols (see the following section, "Labelling Standards"). Labelling on devices must be tested almost as much as the device itself. Obviously, in the case of clinical investigations, where the investigator has more information on the device—its usage, strict techniques to be followed, and so on—the contents of the label and the instructions leaflet

---

**Figure 14.1. Example of a Label for Clinical Investigations
That Meets the Minimal Requirements of Section 13 of
Annex I Without Symbols**

Cardiovascular Catheter—6F, 125 cm length

Catalogue Number:	XYZ123
Contents:	1 unit
Manufacturer	Authorised Representative
Industrial Estate	Commercial Park
Anytown, Out of Europe	Capital City, In Europe

STERILE	Single-use only
Sterilised by Ethylene Oxide gas	LOT: 1995XYZ001
Use by: 1998-07	

Exclusively for clinical investigations

---

are unlikely to be challenged, a separate test of labelling
alone might justifiably be considered by conscientious
manufacturers. However, as mentioned previously, it is
vital to minimise the number of variables in any scientific
experiment, and compliance to the ERs with regard to
labelling will at least minimise variables in this area.

Labelling Standards

Section 13.2 of Annex I clearly states that, where appropriate,
labelled "information should take the form of symbols". These
symbols "must conform to the harmonised standards".\

Harmonised standards are European standards developed
and produced by one of the European standards–making
bodies—CEN (Comité Europeén de Normalisation) or CENELEC
(Comité Europeén de Normalisation Electrotechnique)—under
a mandate from the European Commission. Once these standards have been developed and accepted by the Member States,
their references are published in the *Official Journal of the
European Community*. At this point, they can be said to be

"harmonised". This means that all Member States will adopt these European standards as their own and, in terms of complying with the MDD, these standards take precedence over all other national and international standards.

In the area of device labelling, there are two key standards currently in their final stages of preparation: EN 1041, "Terminology, Symbols, and Information Provided with Medical Devices: Information Supplied by the Manufacturer with Medical Devices"; and EN 980, "Graphical Symbols for Use in the Labelling of Medical Devices". Both of these standards should become harmonised standards in the near future.

The first, EN 1041, deals with the requirements for labelling of medical devices as noted in the MDD with information on specific areas, such as what constitutes a legible label. The second, EN 980, provides the manufacturer with harmonised symbols and graphics to replace multiple translation as appropriate.

EN 1041

At the time of writing, this standard (also true for EN 980) is not yet available in its final form. Therefore, the comments given below must be checked by the reader against EN 1041 as finally published. An important note is that this standard does not set out to specify the language, or languages, to be used for information supplied with a medical device (see "Language Considerations" in this chapter).

The draft EN 1041 has four important normative references:

1. EN 980;

2. EN 60601: Medical electrical equipment (safety requirements: all relevant parts);

3. ISO 31: Quantities and units; and

4. ISO 8601: Data elements and interchange formats— Information interchange and representation of dates and times.

These are all referred to within Section 4 of the standard (Requirements). This section indicates the means by which a manufacturer may meet the ERs of the MDD as noted earlier in this chapter.

The label (or "information supplied") is to be "legible when viewed under an illumination of 215 lux using normal vision . . . at a distance . . ." that is appropriate to the use, size, and shape of the device. If legibility is compromised by the size of the device, the information may be included on an insert, accompanying leaflet, or document.

Other requirements include such items as the following.

- Symbols and reference colours should be used as specified by relevant standards (e.g., EN 980).

- Information supplied by the manufacturer must not be misleading or obscure any other "essential" information.

- The use of SI units for any units of measurement as specified in ISO 31 is recommended, but "other legal units" are allowed. Consideration should be given to ensure that the users of the device understand the information supplied.

- Instructions for use should be on the packaging label, if practical.

- Dates shall be expressed as per ISO 8601. That is, in the format YYYY-MM-DD or similar (e.g., 1995-07-24).

The standard also directs the manufacturer to the use of other standards specific to the devices (as applicable). For example, EN 60601-1, which is relevant for all electromedical devices, and EN 60601-2 (multiple parts) for specific electrical and electronic equipment (e.g., diathermy apparatus). These standards have specific labelling requirements that should be met as appropriate.

EN 980

EN 980 provides manufacturers with harmonised symbols for device labelling. These cover the major elements required by the MDD:

- catalogue or reference number
- sterile

- sterilisation method

- lot number

- serial number

- date of manufacture

- use by date

- do not re-use (or single-use) statement

- see instructions for use

These symbols have been debated and agreed to over a number of years and are acceptable to all Member States. However, language variations in the transposition of the MDD leaves some of the issues relating to this open (see "Language Considerations").

The other issue relating to symbols is the level of education of the user. It is fine to have a system of graphically represented information, but if users do not understand it, then it will be of no value. A programme of education must be established by both the industry and the healthcare professions to ensure that users are cognisant of these symbols. The use of such symbols on clinical investigation devices is a major contribution to this programme, as many investigations are performed at specialist research centres within teaching hospitals. By this means, users will increasingly become aware of the meaning of such symbols on device labels. Whilst not a requirement of either the standard or the MDD, manufacturers may well wish to use both symbols and written statements in parallel for a period of time to ensure that users know, for instance, that a particular device is for single use. This should be considered carefully for reasons of liability, label size, and language issues.

Instructions for Use

The specific requirements for instructions for use that should be included with a medical device are enumerated in Section 13.6, Annex I of the MDD.

- The same information as on the labels (Section 13.3) with the exception of 13.3(d), lot number, and 13.3(e), the use by date.

- Details of the "performances" that the manufacturer intends the product to have. In other words, the details of the procedure in which the device is to be used, the technique to be employed, the pressures it will withstand, flow rates, porosity, and so on. The performance being examined by the clinical investigation should not be included, so as to ensure against a bias in the trial's results.

- Any side effects that may occur as a result of using the device on a patient should be outlined. As part of the risk analysis performed by the manufacturer in order to comply with Annex I, Section 2, there is likely to be some risk that cannot be reduced or minimised in any way, except by informing users of it. The MDD terms this "residual risk" and stipulates that users must be informed about such risk. This will include any side effects that a patient may suffer as a result of treatment or diagnosis when using the medical device in question.

- Details of connections, capability, and functional parameters for the device and any other device or equipment with which it must be used in order to fulfil its intended purpose. As with most of the requirements for instruction leaflets, this information is very likely to be provided in the protocol for the clinical investigation. However, as the device is to be tested in circumstances that mimic as close as possible conditions of normal use, they should still be included with the device.

- Information on the safe installation and correct functioning of the device. Should the device require maintenance and calibration, then this, too, should be included.

- If appropriate, information relating to avoidance of risks associated with the implantation of the device.

- Details of any risks related to interference the device may cause with other specific treatments or diagnoses (e.g., if the device is a piece of electromedical equipment, would it interfere with a pacemaker?).

- Instructions on handling the device should the packaging be damaged or compromised, thus affecting the sterility of the product (i.e., can it be re-sterilised? What technical checks need to be performed to confirm device integrity? What sterilisation method and cycle should be used, if any?).

- Similarly, if the device is designed for multi-use, or re-use, the details of appropriate processing to enable re-use (i.e., cleaning, disinfection, packaging, sterilisation method, number of allowable re-uses, etc.).

- Any information on further treatment that is required prior to the user utilising the device (e.g., pre-clotting of an uncoated vascular prosthesis).

- Details, where applicable, of the radiation that may be emitted from a device, including its nature, type, intensity, and distribution (e.g., radio-therapy equipment).

The instructions for use must also include contra-indications for use of the device and any precautions as necessary. Such information must be sufficient to enable medical staff to inform the patient of these contra-indications and precautions. In particular, the following should be noted:

- Precautions to be taken should the intended performance parameters of the device change or not be met (e.g., inaccurate delivery of drugs from an ambulatory infusion pump).

- Precautions to be taken in relation to electromagnetic fields (e.g., for pacemakers).

- "Adequate information" about any medicinal product (i.e., pharmaceutical) that the device may be designed to deliver. Also, any limitations in the choice of drugs to be delivered should be noted (e.g., with an infusion pump).

- Precautions against contamination or any other unusual risk associated with the disposal of the device after use.

- Should the device incorporate a medicinal product as an integral part, then information on this substance must be provided (e.g., joint prosthesis with an antibiotic component).

- Should the device have a measuring function, the degree of accuracy claimed by the manufacturer must be specified (e.g., for a thermometer or peak flowmeter).

LANGUAGE CONSIDERATIONS

The labelling of devices should, wherever possible, take the form of a symbol or graphic as specified in the appropriate standard. However, the transposition of the MDD into law within individual Member States has led to anomalies relating to the languages that should be used in labelling and instructions for use leaflets. A brief summary of the current status of national requirements is given below:

Austria:	German
Belgium:	Dutch, French, and German
Denmark:	Danish
Finland:	Finnish and Swedish (English may be negotiable in some cases)
France:	French
Germany:	German
Greece:	Greek
Iceland:	Icelandic (for professional use may also use English, Swedish, etc.)
Ireland:	English
Italy:	Italian

Luxembourg:	French and German
Netherlands:	Dutch (English may be negotiable in some cases)
Norway:	Norwegian (English may be used if device is for professional use)
Portugal:	Portuguese
Spain:	Spanish
Sweden:	Swedish (English may be negotiable in some cases)
Switzerland:	French, German, and Italian
United Kingdom:	English on label (any EU language on insert)

As in most cases, these language requirements are part of national laws; non-compliance (certainly in the case of products intended for free sale) will be an illegal act.

Some Competent Authorities, as noted above, will accept that medical professionals using most devices will be trained in relevant procedures and, therefore, the information required in the national language will be minimal. Also, with the increased globalisation of the healthcare industry, the English language is becoming more and more the language of choice for medical conferences. On this basis, some national governments only require labelling and directions for use in their national language if the device is intended for use directly by a patient (e.g., a urological sheath). As these devices are, almost by definition, low risk devices, it is very unlikely that Member States will enforce this aspect of their medical device legislation with respect to devices intended for clinical investigations. However, the previous comments relating to the need for investigations to be performed under conditions as near as possible to the normal conditions of use are still valid here. Labelling of devices for investigational purposes should be in the national language of the user wherever possible.

Multi-Lingual or Country-Specific Labels

In general, there are two options open to the manufacturer who wishes to meet the requirements placed on him by the national language requirements set out in the legislation of Member States:

1. multi-lingual labels and

2. country-specific labels

Both of these solutions have benefits and drawbacks. Some are outlined below.

Multi-Lingual Labels

The use of, for instance, five language labels on medical devices has been a practice for many years. Amongst its positive features is the ability to have one stock of fully labelled, finished products in the warehouse ready for immediate shipment to any country that will accept the languages that are included on the label. For example, a manufacturer may choose the five languages on the label with market size as the guide (e.g., English, French, German, Spanish, and Italian), or strategic usefulness of the language (e.g., French for France, Belgium, Luxembourg, and Switzerland) as the determining factor. The key problem with this multi-lingual approach is the size of the labels for the incorporation of the relevant languages. Some devices are small products with neat packaging (their manufacturers being environmentally conscious) which may not be large enough for five languages, let alone the number required under the various national implementations of the MDD. The increased use of symbols will, of course, reduce this problem, but it is still not the ideal solution, as some 13 languages should be on the label and instructions for use leaflet. This would probably mean having labels twice the size of the product and instruction leaflets that are like small paperback books!

Country-Specific Labels

There is no more straightforward solution to the problem of national language requirements than to supply a device that is labelled specifically for the country in which it is sold. Under this scheme, a manufacturer would hold an inventory of minimally labelled product (i.e., with an identifier and little information) and label it in the national language, or languages, of the country of destination when the customer places the order. This has the advantage of ensuring the labels are clear as there is no clutter from other languages competing for space, and not too small a typeface making it difficult to read the correct language. Also, the customer in Portugal, for instance, will be most impressed that the manufacturer has "produced" a product just for them.

The problems with country-specific labelling, though, include difficulty in stock re-distribution if needed. That is, once labelled for one country, it is very difficult to move excess inventory from one market (e.g., France) to another market (e.g., Greece). This may be an important factor for some companies.

Another problem with this approach is the need for increased handling of the finished product prior to shipment. For example, the labelling operations that would be involved with shipping 10 units to Germany, 5 to Spain, and 20 to the United Kingdom may well increase the amount of product handling over the one multi-lingual label approach. Costs associated with producing 3 types of labels and 3 different operations instead of one means that country-specific labelling is probably not the best solution either—at least not for most medical devices.

Regional Labels

The best solution to the problem of so many languages may be the development of, say, three inventories for the "regions" of Europe. This would be a variant on the multi-lingual labels, with an element of the country-specific concept incorporated as well. Possible groupings could be as follows:

- Swedish, Norwegian, Danish, Finnish, and Icelandic for shipments to Sweden, Norway, Denmark, Finland, and Iceland

- French, German, Dutch, and Italian for shipments to France, Luxembourg, Belgium, The Netherlands, Germany, Austria, Switzerland, and Italy

- English, Spanish, Portuguese, and Greek for shipments to the United Kingdom, Ireland, Spain, Portugal, and Greece

Whilst these considerations may not be absolutely necessary for medical devices intended for clinical investigations, the principle of utilising products in as near to their normal conditions of use as possible still applies. Careful consideration of future strategy should be made prior to using any one language option, especially if the device under investigation is to be used by a non-professional.

REQUIREMENTS OF THE ACTIVE IMPLANTABLE MEDICAL DEVICE DIRECTIVE

The Active Implantable Medical Device Directive (AIMD) (90/385/EEC) has very similar requirements for labelling, instructions for use, and clinical investigations. These are outlined in Article 4, Article 10, Annex I, Annex VI, and Annex VII of the AIMD. The variations are minor. Previous comments for the labelling and instructions for use for medical devices in general apply equally to active implantable medical device labelling and instruction leaflets. As with the comments on the MDD, EN 540, and the other standards referenced earlier, the AIMD should be consulted in conjunction with this chapter.

CONCLUSION

Whilst conformance with the requirements laid down in the MDD is not expressly mandated with regard to the labelling for clinical investigational devices, the foregoing is an attempt to show the reader the indirect requirements of the MDD. Compliance with the MDD at an early stage in the development of a product will speed up the process of CE marking and is as has been shown a requirement under Section 2.2 of Annex X. It will also ensure that the variables that may bias a study are limited and, by this means, contribute to the attainment of the most accurate results possible.

2: *Labelling Requirements in the United States*

As in Europe, the aim of clinical investigations under the Food and Drug Administration requirements is to ensure that devices put into use are safe, efficacious, and perform as intended; that there are no undesirable side-effects related to the use of the product under normal conditions of clinical application. Most of the evidence to demonstrate safety and efficacy (S&E) is found in the device history record (DHR) or device history file (DHF). This information serves as the basis of the information to go into the labelling.

The primary purpose of labelling is that in principle to provide information concerning the product to end user, be that the health-care professional or patient, so that the end-user can use the product safely as directed and intended by the manufacturer. Any claims made in the labelling must be substantiated and valid. Inaccuracies or ommissions could lead to reportable events since they may affect the safe use of the product by health-care professional and/or patient. Liability law lurks around the corners of any piece of labelling.

DEFINITION OF LABELLING

Per the CFR and its guidance documents[1], a label is 'any display of written, printed or graphic matter upon the immediate container of any article . . .' *Labelling*, on the other hand, is defined as 'all labels and other written, printed or graphic matter upon any article or any of its containers or wrappers or

---

1. Labelling: Regulatory Requirements for Medical Devices. Prepared by the Division of Small Manufacturers Assistance and Office of Training and Assistance.

accompanying any article at any time while a device is held for sale after shipment of delivery for shipment in interstate commerce'.

In the United States, labelling is regarded to include

- any information on the product (affixed to the product), and

- any information accompanying the product (e.g., operating manuals or users manuals, advertising materials, brochures, posters, and so forth).

TIP	Whereas the United States requirements refer to 'operating manuals' or 'users manuals', the term used in the European requirements is 'instructions for use'. User manuals, could also be used to refer to patient information where the product is intended to be provided to a patient for use.

It could be argued that advertising is not considered 'labelling' as such. According to United States jurisprudence most advertising is considered to be 'labelling'.

TIP	Advertising materials are promotional material. They may include sales brochures, advertisements in journals, sales materials for hand-outs at congresses, trade shows, professional meetings, etc. Under the European requirements, advertising is not considered to be labelling *per se.* Promotional materials or advertising materials nonetheless should be subject to review for content and accuracy of statements. These materials are subject to liability laws in the same manner as the labelling.

TIP	In some countries advertising materials or promotional materials are subject to approval by appropriate regulatory authorities or advertising counsels.

UNITED STATES LABELLING REQUIREMENTS

There are two main elements in the Code of Federal Regulations (CRF):

- Quality System requirements for labelling and control thereof, and

- Actual requirements to appear on product or accompany a product.

Quality System Requirements

Per the United States QSR, each manufacturer needs to have in place procedures (i.e., Standard Operating Procedures) for labelling including how labelling activities are

- defined

- reviewed

- controlled, and

- in case of translations, how the translations are handled, reviewed and controlled.

Furthermore, labels and labelling must be

1. printed and be legible at all times;

2. subject to inspection prior to release for use on production floor.

3. identified properly so that storage is easy and mix-ups are prevented.

4. have a control number identifier for traceability purposes.

 Control number identifier could be, for example, a part number or a date and revision level. Key is that the labelling materials need to be rapidly identified and traceable. Use of a date and revision level is recommended to meet both United States and European requirements.

These elements are similar to those defined and implied in the Quality System requirements as set out by ISO 9000/EN 46000 series.

TIP	Labelling is subject to design, review, control and release. All elements therefore need to be defined in the appropriate procedure(s).

ACTUAL LABELLING REQUIREMENTS

There are a number of key characteristics for United States labelling. These can be summarised as follows:

- False and misleading statements are prohibited and is considered to be misbranding.

- All labelling statements must have be prominent and conspicious whereby:

    - they are to appear on all labelling

    - extend the required labelling over the entire width and breadth of the packaging

    - not be obscured by non-required labelling

    - must have a sufficient type-set rending it readable

TIP	Specific requirements pertaining to font size and type are found in the CFR. Ensure that labelling (i.e., information on the pouch, device or packaging as well as in the instructions for use (operating or user manuals) are printed in a font which corresponds to the United States requirements.

Further to the above, the CFR defines specific labelling requirements in two main levels: general and specific.

General

According to the CFR, all products, whether commercialised or not, need to have the following items included in the labelling:

- Manufacturer's or distributor's name, city, state and post-code;

- When the product is manufactured by a different person than that identified on the labelling; the phrase 'manufactured for . . ./distributed by . . .' is to be used

- Statements concerning:

  1. the intended use of the devic

  2. the purpose for which the product can be used

  3. the conditions or purpose under which the product can be used

- Frequency of re-use

- Method of application

- Cautionary statement, i.e. the phrase 'Federal law restricts the distribution of this device to sale or on the order of a physician'.

 Under the Food and Drug Administration 1997 Modernisation Act, section 126 section 503 (b)(4) of the FDC Act is to be amended to eliminate the requirements for inclusion of the label statement 'Caution: Federal law' statement. The new requirement is the replacement of the text, at a minimum, with the symbol ℞.

Specific

In addition to the above, the requirements also define a set of secondary requirements which are to be appear on the label in case of specific situations:

- For devices under clinical investigation, the labelling is to include:

  1. Statement 'Caution: Investigational device. Limited by Federal United States law to investigational use'.

 Devices which subject to approval (i.e., 510k clearance or PMA/IDE) require to have the above mentioned statement prominently displayed when used in the clinical investigation. A draft instructions for use and label may be useable provided Food and Drug Administration and Institutional Review Board agree.

| TIP | For devices that are all ready on the market, which are involved in clinical investigations for reasons other than safety, efficacy or performance need not include a phrase 'investigational device'. All labelling requirements for commercialised devices do need to be complied with as stipulated by the approval for the device. |

---

    2. Contra-indications

    3. Any risks associated with the device

    4. A list of potential adverse events

    5. Statements concerning any interferences which the other substances or devices which may be caused by the product

    6. All necessary and appropriate warnings and precautions.

- For products which are sterile, further additional requirements are needed:

    1. Information or statements pertaining to special cleaning methods as appropriate.

    2. Statement 'Caution: sterile device' or similar statement.

    3. Any statements concerning resterilisation and appropriate methodologies.

    4. Statement that the device is for single use only.

    5. Warnings and precautions concerning sterility, as appropriate.

    6. Shelf-life date or expiry date.

| TIP | Obtain information from the appropriate Food and Drug Administration Division at CDRH if guidance documents for a particular device are available. Alternatively, one may contact DSMA. |

The following tips need to be taken into consideration:

- When a device is under clinical investigation or an experimental product, for which a 510k or IDE or PMA may be pending, be displayed at trade-fairs or congresses, these products must be labelled 'not for commercial purposes/under approval' or as 'work in progress'.

- For products which are not yet approved in the European Union, for example, those that do not bear a CE mark, a similar label should appear where possible.

- For devices which are not experimental but for which an approval has been requested at FDA, a manufacturer should be also label them 'under approval/not for commercial use'.

TRANSLATIONS

The United States requirements specify that all labelling must be in English, 'with the exception of products distributed solely within Puerto Rico or a United States territory where the predominant language is other than English'.

A United States based manufacturer shipping products into regions of the world where the regulatory requirements specify labelling to be in the language of the designated market-place, is required to 'ensure that all representations made in the original labelling are also made in the foreign language', i.e., everything on the label requires to be translated as it is written.

The following tips and hints should be taken into consideration when involved in labelling and the translations thereof:

- Define the process of labelling and translations in standard operating procedures (i.e., who does what when and how).

- Ensure that translations are done by accredited and qualified translation agencies.

> | TIP | For labelling developed in the United States, it is recommended to use European based translation agencies or agencies who can ensure translations are done in the country of the language required, where possible. Authorities worldwide reserve the right to review translations for correct use of the national language.

- Ensure that translations are reviewed by field personnel and the regulatory function. This review takes place before translation on the original text and after translation. Such review ensures the correct terminology has been used

- Ensure any translators are trained. Document the training.

- Ensure review process is documented in some way or other.

- Ensure that translations are accompanied by a certificate of authenticity from the translation agency.

WHAT GOES WHERE

When examining labelling in general, it is important to determine what goes on the pouch label, on the device, and in the instructions for use or accompanying instruction leaflet.

> | TIP | Most pertinent information which allows safe use and traceability of the product should always appear on the label proper. Remaining information should go into the instructions for use or accompanying instructions leaflet. Most labels are insufficiently large to allow all information to appear on them as such. Common sense should prevail.

In the European requirements Annex I of the Directives specifies that labelling on the pouch or packaging or device should include all the information as defined in the Annex I section *13* *'as far as practical and appropriate'*. This implies that pertinent information may appear on the (pouch) label to allow rapid identification of contents, traceability, storage and product use. The remaining information can be placed in the instructions for use. The same is inferred in essence by the United States requirements.

Additional labelling requirements are influenced by such factors as, for example, the submission of a PMA or 510-k, establishment inspections, clinical data results, any product specific guidances issued by Food and Drug Administration and so forth.

Designing and developing the labelling to accurately reflect the claims and safety profile of the product is essential for both the United States regulatory systems and European regulatory systems. Although overall the requirements are similar the content (i.e., wording versus symbology) become critical issues. For instance, symbols widely accepted by other nations are not acceptable to the United States regulatory authorities. Manufacturers face the perspective of generating country-specific labelling in many cases as expounded in part 1. Table 14.1 gives a comparative overview of what is required by United States and European regulatory requirements to appear on the labelling. The table may be a useful guideline for manufacturers faced with developing suitable labelling in order to meet the needs of both regulatory systems.

TABLE 14.1. WHAT GOES WHERE FOR UNITED STATES AND EUROPEAN COMPLIANCE

NOTE Pertinent items which need to appear on the pouch or package label or on the device are marked with a *.

NOTE Parenthesis around the (3) indicate not specifically specified in the labelling requirements but recommended to be applied.

ITEM	UNITED STATES		EUROPEAN	
	LABEL	IFU	LABEL	IFU
MANUFACTURER INFORMATION:				
Manufacturer Information*	✓	✓	✓	✓

NOTE For the European requirements: name/address and/or logo of the manufacturer. For the United States, name, city, state, post-code is sufficient or name and telephone number.

ITEM	UNITED STATES		EUROPEAN	
Distributor identification*	✓	✓		

NOTE In some country transposition texts of the Directives, the name of the distributor or importer may also be necessary. For the United States, name, city, state, post-code is sufficient or name and telephone number.

ITEM	UNITED STATES		EUROPEAN	
Name and address of authorised representative.*			✓	✓

NOTE The European requirements state: manufacturer and/or person responsible for product and/or authorised representative.

TIP Include the name of the person responsible for the product on the label or the authorised representative. In some cases, an AR may not wish to have his or her name on the labels; AR do not take responsibility for the product having solely a regulatory function.

ITEM	UNITED STATES		EUROPEAN	
Use of statement 'manufactured for . . . / distributed by . . .' where appropriate.*	✓			

Table 14.1 continued on next page.

Table 14.1 continued from previous page.

	UNITED STATES		EUROPEAN	
ITEM	**LABEL**	**IFU**	**LABEL**	**IFU**
PRODUCT IDENTIFICATION INFORMATION				
Name of device and catalogue number, if applicable.*	✓			✓
In both sets of requirements, all accessories (cq detachables or components), accompanying the product must be identified.	✓	✓	✓[2]	✓[3]
Sterility statement, in case of sterile products.* For the USA, the phrase to be used is 'Caution: sterile device' although other phraseologies may be accepted.	✓	✓	✓[3]	
Lot number or serial number.*	✓	✓	✓	
Use by date (European phrase)* OR Shelf-life or expiry date (USA phrase), as appropriate.*	✓	✓	✓[4]	✓[5]
Statement that the device is for single use.*	✓	✓		
Warnings and precautions concerning sterility as appropriate.	✓	✓		
Any special storage and/or handling conditions.	✓		✓[6]	✓

NOTE

2. The word 'STERILE' along with method of sterilisation is required for Europe using the appropriate symbol. The symbol to be used for the label while the phraseology to be used in the instructions for use.

3. The word 'LOT' or 'SN' for batch code or serial number per symbol as defined in the standards. See Part 1.

4. The use of the appropriate symbol.

5. Symbol on the pouch or packaging; statement 'Sterile Device: Sterilised by [method]' in instructions for use.

Table 14.1 continued on next page.

Table 14.1 continued from previous page.

ITEM	UNITED STATES		EUROPEAN	
	LABEL	IFU	LABEL	IFU
Any special operating instructions NOTE: See footnote[6]			✓	✓
Any warnings and/or precautions NOTE: See footnote[6]	✓	✓	✓	✓
Revision control identifier or code of the label or labelling[7]	✓	✓	✓	✓
Intended purpose.	✓	✓	(✓)	(✓)
SPECIFIC INFORMATION OR PHRASEOLOGY				
Year of manufacture for active devices. This *may be* part of the *SN*; it may appear only on the device. The reader does well to consult with their respective Notified Bodies.	✓	✓	(✓)[8]	✓
As appropriate, the phrase 'custom made device'.			✓	✓
As appropriate, the phrase 'exclusively for clinical investigations'.			✓	✓

6. For Europe the use of the appropriate symbol directing the user to the instructions for use is appropriate and acceptable.

7. Under the quality system requirements, labelling is considered to be a controlled document. Although specifically stated in the United States FDA requirements, applicable is also to compliance under the European requirements, particularly, those defined in the quality system standards.

8. For Europe this would only apply to the pouch label if it is not obvious to the user from the package what the intended purpose of the product is.

Table 14.1 continued on next page.

Table 14.1 continued from previous page.

ITEM	UNITED STATES		EUROPEAN	
	LABEL	IFU	LABEL	IFU
As applicable, the phrase 'Caution: Investigational Device: Limited by Federal United States law to investigational use'.	✓	✓		
Phrase: 'Federal law restricts this device to sale by or on the order of of a physician'.	✓	✓		
At a minimum this may be replaced, under the 1997 Modernisation Act by the prescription symbol.				
DESCRIPTIVE INFORMATION				
Risk Information: Any precautions and warnings concerning risks inherent in the product.		✓		✓
List of potential adverse events or undesirable side-effects.		✓		✓
Instructions about how the device is to be connected to other device, as appropriate and applicable.		✓		✓
Information about the effects of any interferences on the device, i.e., effects of other materials, substances or devices on the product.		✓		✓
Description of the Product.		✓		✓
Frequency of use of the product (i.e. single use; multiple use).		✓		✓
As appropriate and applicable, instructions about the reusability of the product, (i.e. the number of re-uses, definition of processes for disinfection or special cleaning methods or resterilisation).		✓		✓
Where appropriate, in cases of resterilisation, the details of the resterilisation.		✓		✓

NOTE

Table 14.1 continued on next page.

303

Table 14.1 continued from previous page.

ITEM	UNITED STATES		EUROPEAN	
	LABEL	IFU	LABEL	IFU
Duration of use of the product.		✓		
Method of application.		✓		
All information to use the device safely.		✓		✓
Any installation or implantation instructions, as appropriate and applicable.		✓		✓
Maintenance and calibration information, as appropriate.		✓		✓
Information to avoid certain risks in connection with implantation of the device, where applicable and as appropriate.				✓
Phrase pertaining to what to do in case of damaged package or damaged device. The standard accepted phrase is 'If package is damaged, then do not use' or something similar.	✓		✓	✓
Indications		✓		✓
Contra-indications		✓		✓
Details of any processing activity the device needs to undergo prior to use by the user.		✓		✓
Where appropriate, for emitting devices, details on the intensity and distribution of radiation, radiology, or ionising energy.		✓		✓

Table 14.1 continued on next page.

Table 14.1 continued from previous page.

ITEM	UNITED STATES		EUROPEAN	
	LABEL	IFU	LABEL	IFU
Identification of any medicinal substances incorporated into the device as an integral part of the product.				✓
Clarification of any symbols used should they be self-designed. For recognised symbols from the standards such clarification may not be necessary.		✓		✓
Specific regulatory information concerning the device as defined in any specific product documents, Food and Drug Administration regulations, or (international or national) standards.[9]				

9. This requirement emerges from Food and Drug Administration labelling requirements.

15

Ethics Committee Approval

Herman Pieterse
Profess® Medical Consultancy BV
Heerhugowaard, The Netherlands

The Code of Federal Regulations (CFR) of the U.S. Food and Drug Administration (FDA), and the device directives amplified by EN 540 define the responsibilities, composition, functions and operations, and procedures for Ethics Committees (ECs) or Institutional Review Boards (IRBs). These guidelines have to be followed in order to arrange for and obtain a proper ECs/IRB approval for the execution of a clinical trial.

Although clinical research managers and study monitors are convinced that the primary responsibility of ECs/IRBs is to safeguard the rights, safety, and well-being of all trial subjects, they criticise these institutions as the major cause of delay in setting up clinical trials. The major reasons behind these delays are the random constitutions of the committees, inadequate training, different review methods, infrequent meeting dates, protocol waiting lists, and so on.

The objective of this chapter is to provide the reader with instructions on how to avoid problems, or at best how to deal with the problems as smoothly and quickly as possible. First, some background information is given with respect to the current requirements for the responsibilities of ECs/IRBs, the composition of the committee, its functions and operations, and procedures.

THE RESPONSIBILITY OF ETHICS COMMITTEES

The purpose of an EC/IRB is to safeguard the rights, safety, and well-being of all trial subjects, with special attention given to trials that may include vulnerable subjects, taking into account the scientific procedure and the concerns of the local community. ECs/IRBs should provide timely, comprehensive, and independent ethical reviews of the proposed studies according to the Declaration of Helsinki and Good Clinical Practice (GCP) regulations. They are responsible for acting with due regard according to the requirements of relevant regulatory agencies, applicable laws in the respective countries, and in good faith with respect to both the applicants and the community.

COMPOSITION, FUNCTIONS, AND OPERATIONS

The EC/IRB should consist of a reasonable number of members, who collectively have the qualifications and experience to review and evaluate the science, medical aspects, and ethics of the proposed trial. It is recommended that the EC/IRB include at least

- five members;
- one member whose primary area of interest is in a non-scientific area; and
- one member who is independent of the institution/trial site.

Only those EC/IRB members who are independent of the investigator and the sponsor of the trial should vote/provide an opinion on a trial-related matter. A list of EC/IRB members and their qualifications should be maintained.

The EC/IRB should perform its functions according to written operating procedures; should maintain written records of its activities and minutes of its meetings; and should comply with the GCPs and with the applicable regulatory requirement(s). Decisions should be made at announced meetings at which at least a quorum, as stipulated in its written operating procedures, is present. Only members who participate in the review and discussion should vote/provide their opinion and/or advise.

The investigator may provide information on any aspect of the trial, but should not participate in the deliberations of the EC/IRB or in its vote/opinion. In some cases it is advised to invite the writer of the clinical investigation plan, often a representative of the sponsor, to the meeting of the EC/IRB because this person is the true expert with respect to the objective, purpose, and design of the study protocol. Non-members may be invited with expertise in special areas for assistance.

The EC/IRB should obtain

- the trial protocol(s)/amendment(s);

- written informed consent form(s) and consent form updates that the investigator proposes to use in the trial;

- subject recruitment procedures (e.g., advertisements);

- any written information to be provided to subjects;

- the Investigator's Brochure, or Report of Prior Investigations;

- available safety information;

- information about payments and compensation available to subjects (no-fault insurance policies);

- the investigator's current curriculum vitae and/or other documentation that shows his/her qualifications; and

- any other documents that may be needed to fulfil its responsibilities.

Indemnity agreements between the sponsor and the investigator are essential documents required by most ECs/IRBs. Insurance certificates could also be essential documents as assurance that risk liability (tort liability) has been covered by the sponsor. Most ECs or IRBs today use application forms that require a lot of information.

| TIP | It is recommended that a diagrammatic representation ("flowchart") of the protocol be submitted. It may help the committee gain a quick and comprehensive overview of the experiment. Such a flowchart or time/measurement schedule is, in most cases, an integral part of the study protocol. |

The EC/IRB should design and publish guidelines for the submission of an application for the ethical review of proposed biomedical research. These guidelines should include the

- name(s) and address(es) of the EC/IRB members to whom the application material is to be submitted;

- number of copies to be submitted;

- language(s) in which documents are to be submitted;

- required application form(s);

- required documentation;

- required format;

- deadlines for review dates; and

- means by which applicants will be informed of incompleteness.

Operations of the EC/IRB

The EC/IRB should review a proposed clinical trial within a reasonable period of time and document its views in writing, clearly identifying the trial, the documents reviewed, and the dates for

- approval/favourable opinion;

- modifications required prior to its approval/favourable opinion;

- disapproval/negative opinion; and/or

- termination/suspension of any prior approval/favourable opinion.

A reasonable period of time for the approval process is **six to eight weeks.** In some countries, it is much longer than eight weeks. However, once the EC/IRB restructures its operations and standardises submission by issuing standard application formats, this approval period will become more feasible.

The EC/IRB should consider the qualifications of the investigator for the proposed trial, as documented by a current curriculum vitae and/or by any other relevant documentation requested by the committee.

The EC/IRB should conduct a continuing review of each ongoing trial at intervals appropriate to the degree of risk to which the trial exposes human subjects—but at least once per year. In practice, however, continuing review is only performed by IRBs in the United States, not in Europe. Most of the ECs/IRBs in Europe are currently voluntary organisations; but they will be organised more professionally in the near future. Then these committees will have more time to be actively involved during the execution period of the clinical trial. The opinion of most of the ECs/IRBs is that continuing review should be the responsibility of the investigator and the sponsor. The ECs/IRBs should be informed when, for instance, the progress is in jeopardy, but they do not wish to take action in this respect.

The EC/IRB should review both the amount and method of payment to subjects to ensure that neither presents problems of coercion or undue influence on the trial subjects. Payments to a subject should be pro-rated and not wholly contingent on the subject's completion of the trial. The ECs/IRBs should ensure that information regarding payment to subjects, including the methods, amounts, and schedule of payment to trial subjects, is set forth in the informed consent form and in any other written information provided to the subjects. Pro-rations of payments should be specified.

Special Situations

- The EC/IRB may request that more information be given to subjects when, in their judgement, the additional information would add meaningfully to the protection of the rights, safety, and/or well-being of the subjects.

- When a non-therapeutic trial is to be carried out with the consent of the subject's legal representative, the EC/IRB should determine that the proposed protocol and/or other document(s) adequately address relevant ethical concerns and meet applicable regulatory requirements for such trials.

- Where the protocol indicates that prior consent of the trial subject or the subject's legal representative is not possible, the EC/IRB should determine that the proposed protocol and/or other document(s) adequately address relevant ethical concerns and meet applicable regulatory requirements for such trials (i.e., in emergency situations).

Procedures of the EC or IRB

The EC/IRB should establish, document in writing, and obey the following procedures:

- determine the composition (i.e., names and qualifications of the members) and the authority under which it is established;

- schedule, notify its members of, and conduct its meetings;

- conduct initial and continuing review of trials;

- determine the frequency of continuing reviews, as appropriate;

- provide, according to the applicable regulatory requirements, expedited review and approval/favourable opinion of minor change(s) in ongoing trials that already have received approval/favourable opinion. The term *minor* should be defined.

- Specify that no subject should be admitted to a trial before the EC/IRB issues its written approval/favourable opinion of the trial;

- Specify that no deviations from, or changes to, the protocol should be initiated without prior written EC/IRB approval/favourable opinion of an appropriate amendment, except when necessary to eliminate immediate hazards to the subjects or when the change(s) involve only logistical or administrative aspects of the trial (e.g., change of monitor(s), telephone number(s);

- Specify that the investigator should promptly report to the EC/IRB

  - deviations from, or changes to, the protocol to eliminate immediate hazards to trial subjects;

  - changes to the protocol increasing the risk to subjects and/or significantly affecting the conduct of the trial;

  - all adverse incidents that are unexpected or unanticipated (see chapter 23); and

  - new information that may adversely affect the safety of subjects or the conduct of the trial;

- Ensure that the EC/IRB promptly notifies the investigator/institution in writing concerning

  - its trial-related decisions/opinions;

  - the reasons for its decisions/opinions; and

  - procedures for appeal of its decisions/opinions.

The EC/IRB should retain all relevant records, such as written procedures, membership lists, lists of occupations/affiliations of members, submitted documents, minutes of meetings, and correspondence, for a pre-defined period (e.g. at least three years) after the completion of a trial, and it should make these records available on request from the regulatory authority(ies). The EC/IRB may be asked by investigators, sponsors, or regulatory authorities to provide its written procedures and membership lists.

ECs/IRBs should standardise their submissions by issuing a standard application form. Table 15.1 is an example of such a standard form.

Table 15.1. Standard Application Form for the EC or IRB

No.	Section	Questions	Answer for this Study
1	General information	Title of the investigation	
		Trial code	
		Planning trial	
		Anticipated start date (start of subject inclusion)	
		Anticipated completion date (reporting)	
2	Investigator(s)	Sponsor name and address	
		Name of the liaison of the sponsor and function	
		Name and affiliation of principal investigator	
		Number of planned investigators, with affiliations	
3	Investigational product	Regulatory status (with details on intended purpose)	
		Details on the function and performance of the device	
4	Rationale	Primary objective of the investigation	
		Purpose of the investigation. (Describe concisely the clinical relevance, risks, and benefits for the intended study population in relation to the intended purpose of the device.)	
5	Methodology	Design of the study, including inclusion period, number of treatment arms, controls, and treatment/observation period	

Table 15.1 continued on next page.

Table 15.1 continued from previous page.

No.	Section	Questions	Answer for this Study
6	Liability	Number of subjects to be included in the investigation	
		Number of investigational centres to be recruited, with the anticipated number of subjects per centre	
		Does the sponsor indemnify the investigator for risk (tort) liability caused by the experiment?	
		Submit an indemnity statement or insurance certificate	
7	Subjects	How will the subjects be recruited?	
		Inclusion criteria (see protocol page ___)	
		Exclusion criteria (see protocol page ___)	
		Inconvenience for the subject	
		• Psychic load	
		• Physical load	
		Which interventions will be performed?	
		benefit/risk ratio	
		• Total duration of the investigation	
		• Risks for the subjects (other than side effects)	
		• Benefits for the subjects	

Table 15.1 continued on next page.

Table 15.1 continued from previous page.

No.	Section	Questions	Answer for this Study
8	Privacy protection	• Precautions to avoid and limit risks • Ratio between benefit and risks Access to patient file by third parties Which persons will obtain direct access? Did these persons sign a confidentiality agreement? Are these persons qualified?	
9	Informed consent/ subject information	How will informed consent be obtained? How will the subjects be informed? What is the planned time-frame for deciding on participation? Inclusion of the informed consent Inclusion of subject information sheet	
10	Financial consequences	What is the fee for the investigators (hot issue)? What is the compensation fee for the subjects?	
	Name and affiliation of the applicant:		

WHAT IS THE STUDY MANAGER'S PART IN OPTIMISING THE PREPARATION PROCESS?

The study manager or study monitor should contact the EC/IRB as early on as possible during the site selection process. This could even be as soon as the investigator has expressed an interest in the trial or at least has agreed to participate.

The study manager should find out the following.

- Is the EC/IRB properly constituted?

- When does the EC/IRB meet and how frequently?

- Are there special requirements regarding consent forms, mandatory NCR (non-carbon required) copies, and so on?

- Should a standard application form be completed?

- How many copies have to be submitted? Specify for each document to be submitted.

- How long before the meeting should the documentation be submitted?

- Is there a protocol waiting list?

- Are there any special indemnification requirements?

- How soon after the meeting is a response normally issued by the EC/IRB?

- What administration fees should be paid and to what account?

'IP   It is recommended that within the company (at least) an in-house file containing the information listed previously be compiled for all ECs/IRBs for future reference in order to prevent reinventing the wheel each time.

Although the investigator is primarily responsible for submission of the study proposal to the EC/IRB, the study manager should take the lead to ensure that all the required documentation is received by the EC/IRB as soon as possible for consideration at the next meeting. Do not wait until the last day. Murphy's Law is always applicable in these situations. Submissions received on the last acceptable day before a meeting are more likely to be delayed.

Sometimes, the investigator will attend the meeting of the EC/IRB. The study manager should then ensure that all members are well informed and briefed on the protocol.

Although the study protocol is an essential document for judging the scientific merit of an experiment, many of the comments received from EC/IRBs relate to consent forms and patient information.

| TIP | Templates and/or standard informed consent forms are widely used in the industry. Encourage the investigator's input on this form because the investigator best knows how to word certain information to his/her patients. Moreover, investigators are often more familiar with local requirements and nuances. |

The day of the meeting of the EC/IRB is often beyond the control of the investigator and the sponsor. However, a number of ECs/IRBs welcome the direct input of either the investigator and staff and/or the sponsor company at the meeting. Especially, the author of the protocol could contribute meaningful information to the discussion of the protocol (i.e., why has this been included in the protocol?). Some EC/IRBs do not have a sufficiently broad makeup to encompass the range of experience and scientific expertise required by a protocol.

| TIP | The presentation of the study protocol by either the author of the protocol or the investigator can overcome costly delays and has proven helpful to those sponsor companies who have tried this approach. |

| TIP | It is worthwhile investigating if a particular EC/IRB accepts or considers the approval of other IECs/IRBs in cases of a multi-centre trial. Frequently, ECs/IRBs of general hospitals at least partially accept the approval from committees based in academic hospitals. The study manager should investigate this factor before submitting documentation to an EC or IRB. |

WHAT HAPPENS AFTER THE STUDY PROPOSAL HAS BEEN APPROVED BY THE EC/IRB?

A number of ECs and IRBs have administrative struggles. All too often, the secretary of the EC/IRB phones the investigator the day after their meeting to inform him/her on the outcome. However, the written confirmation comes two weeks later. When the letter stating the EC's or IRB's approval arrives, the study manager should check it for the following.

- Does the letter specify which protocol has been approved (protocol title, company name, version of the protocol, and date)?

- Does the letter specify which consent form has been approved (version number and date)?

- Has the list of members present at the meeting been included?

- Are the name and address of the EC/IRB clearly mentioned in the letter?

- Has the letter been duly signed and is the signature clear (i.e., is the name of the person who signed the letter mentioned)?

The study manager should plan for pitfalls beforehand by giving clear and concise information to the IEC/IRB—even by diplomatically discussing with the administrator of the EC/IRB which information should appear on the approval letter.

 TIP   A checklist or EC/IRB review form could be sent to the administrator of the IEC/IRB, documenting all of the items contained in the submission (and their dates) and in which the chairman simply ticks off his/her approval and adds any relevant comments.

APPENDIX A. COMPARISON OF EUROPEAN AND U.S. GCP REQUIREMENTS

Responsibility/Subject	FDA Regulations	European GCP
EC/IRB membership	At least five members with varying backgrounds, one of which is non-scientific; no conflicting interests	At least five members, one of whom is non-scientific and one of whom is independent of investigator, sponsor, and trial site
EC/IRB operations	Written procedures required	Written procedures required
EC/IRB continuing review	Yes	Yes
EC/IRB records	Full records required	Full records required
Signing protocol	Sponsor's obligation; sponsor to ensure that investigator complies with protocol; no signatures required	Investigator to sign the protocol
Protocol amendments	Notification to IRB of major amendments with approval (i.e., changes that will have a significant effect on the safety of subjects)	Major amendments to the EC for approval
Selection of qualified investigators	Sponsor should select investigator	Sponsor to select the investigators

Appendix A continued on next page.

Appendix A continued from previous page.

Responsibility/Subject	FDA Regulations	European GCP
GCP noncompliance by investigator	Sponsor must secure compliance or discontinue shipment of product, and close down the centre	Sponsor must secure compliance or discontinue shipment of products, and close down the centre
Device accountability	Sponsor must keep adequate records of all shipments and receipts of products	Investigator and sponsor both are responsible for adequate record-keeping of allocated/used devices
Investigator's Brochure	Sponsor's responsibility to investigator and gives sufficient time for review	Sponsor sends updates
Estimation of recruitment rate	Monitor's responsibility	Investigator should demonstrate a potential for recruiting the
	required number of suitable subjects within the agreed recruitment period	
Investigator statement		Signed protocol
Appoint study coordinator	Required; appointed by investigator	Investigator should have adequate number of qualified staff

Appendix A continued on next page.

Appendix A continued from previous page.

Responsibility/Subject	FDA Regulations	European GCP
Retention of documentation	Depending on the type of device, 2 years after approval of 5 years from delivery of data to FDA.	For an appropriate time (i.e. period of at least 15 years) however related to lifetime of product as established by manufacturer (e.g. for implants, retain collected records forever.)
Information to study personnel	Investigator's obligation	Investigator's obligation
Reporting results	Investigator's obligation	Investigator's obligation to submit status; promptly written report on any problems affecting the conduct of the trial required
Approval of final report	Responsibility of sponsor	Responsibility of sponsor
SOP and internal audit	Sponsor's obligation for extended quality system	Sponsor's obligation; internal auditing mandatory
Appointing monitor	Monitor must be qualified	Sponsor to appoint qualified monitors and continue training
Data management and publication policy	No regulations	Sponsor's obligation

Appendix A continued on next page.

Appendix A continued from previous page.

Responsibility/Subject	FDA Regulations	European GCP
Compensation for patients	No requirements; should be described in informed consent form	Required even though national laws may not be specific
Indemnification for investigators	Not required	Required by medical liability law
Source documents verification	Direct access required	Direct access required
If applicable, nontherapeutic trials in legally restricted persons	To be decided by IRB	This is generally prohibited by law except if the EC approves such a trial when no alternative target group of subjects can be used and risk/inconvenience is minimal

APPENDIX B. CHECKLIST FOR EC/IRB APPROVAL

1. Collect Information

   a. Contact person at EC/IRB:

      _____

   b. Documentation to be submitted

      • Study protocol and amendments ☐

      • Information to be given to the subjects ☐

      • Informed consent ☐

      • Procedure for recruitment of subjects ☐

      • Investigator's Brochure, including
        available safety information ☐

      • Information on anticipated pro-rated
        payment to the subject for participation
        in the trial and the anticipated expenses ☐

      • Current curriculum vitae of the
        investigator and/or other documentation
        ensuring the suitability of the
        investigator ☐

      • Insurance certificate ☐

      • Any other information deemed necessary

        Specify: _____ ☐

   c. To whom should the documentation be sent?

      _____

      • Send at the latest on _____(date)

      • Should a specific application form be
        completed? yes ☐ no ☐

      • If yes, who could supply this form?

        _____

d. Review procedure:

- Meeting agenda:

  First meeting _____ (date)

  Next meeting thereafter _____ (date)

- Is presence of investigator allowed
  at meeting?   yes ☐ no ☐

- Is presence of author of protocol
  allowed at meeting (if he/she is an
  employee of the sponsor)?   yes ☐ no ☐

- When is written judgement/approval
  expected? _____ (date)

e. Are other approvals required prior
to the start of the trial?   yes ☐ no ☐

  If yes,

- Approval by the director of patient
  care?   yes ☐ no ☐

- Approval by the director of the
  hospital's financial department?   yes ☐ no ☐

- Any other approvals required?   yes ☐ no ☐

  If yes, specify: _____

2. Approval Study Proposal by the EC/IRB:

- Submission date of required documentation

  _____ (date)

- Meeting date of EC/IRB to judge the project

  _____ (date)

- Date of receipt of written EC/IRB approval

  _____ (date)

- Approval/favourable opinion of the EC/IRB:

Negative: If the EC/IRB has any questions or remarks, respond to the queries as soon as possible before the date of the next meeting. Appeal of the IEC's/IRB's decision should be arranged according to the applicable law.

Positive: Request a written approval indicating the

- address of the EC/IRB;
- composition of the EC/IRB;
- specific documents that have been approved— title, identification, and dates of respectively the
    - study protocol;
    - informed consent;
    - information to be given to the subject; and
    - information on recruitment procedures; and
- name and address of the applicant; and
- date of the approval.

When all required approvals have been received on paper, then **START THE INVESTIGATION.**

APPENDIX C. DECLARATION OF HELSINKI

Recommendations Guiding Physicians in
Biomedical Research Involving Human Subjects
Adopted by the 18th World Medical Assembly
Helsinki, Finland, June 1964

and amended by the
29th World Medical Assembly
Tokyo, Japan, October 1975

35th World Medical Assembly
Venice, Italy, October 1983

41st World Medical Assembly
Hong Kong, September 1989

and the
48th General Assembly
Somerset West, Republic of South Africa
October 1996

Introduction

It is the mission of the physician to safeguard the health of the people. His or her knowledge and conscience are dedicated to the fulfillment of this mission.

The Declaration of Geneva of the World Medical Association binds the physician with the words, "The health of my patient will be my first consideration", and the International Code of Medical Ethics declares that, "A physician shall act only in the patient's interest when providing medical care which might have the effect of weakening the physical and mental condition of the patient".

The purpose of biomedical research involving human subjects must be to improve diagnostic, therapeutic, and prophylactic procedures and the understanding of the aetiology and pathogenesis of disease.

In current medical practice most diagnostic, therapeutic, or prophylactic procedures involve hazards. This applies especially to biomedical research.

Medical progress is based on research which ultimately must rest in part on experimentation involving human subjects.

In the field of biomedical research a fundamental distinction must be recognized between medical research in which the aim is essentially diagnostic or therapeutic for a patient, and medical research, the essential object of which is purely scientific and without implying direct diagnostic or therapeutic value to the person subjected to the research.

Special caution must be exercised in the conduct of research which may affect the environment, and the welfare of animals used for research must be respected.

Because it is essential that the results of laboratory experiments be applied to human beings to further scientific knowledge and to help suffering humanity, the World Medical Association has prepared the following recommendations as a guide to every physician in biomedical research involving human subjects. They should be kept under review in the future. It must be stressed that the standards as drafted are only a guide to physicians all over the world. Physicians are not relieved from criminal, civil, and ethical responsibilities under the laws of their own countries.

I. Basic Principles

1. Biomedical research involving human subjects must conform to generally accepted scientific principles and should be based on adequately performed laboratory and animal experimentation and on a thorough knowledge of the scientific literature.

2. The design and performance of each experimental procedure involving human subjects should be clearly formulated in an experimental protocol which should be transmitted for consideration, comment, and guidance to a specially appointed committee independent of the investigator and the sponsor provided that this independent committee is in conformity with the laws and regulations of the country in which the research experiment is performed.

3. Biomedical research involving human subjects should be conducted only by scientifically qualified persons and under the supervision of a clinically competent medical person. The responsibility for the human subject must always rest with a medically qualified person and never rest on the subject of the research, even though the subject has given his or her consent.

4. Biomedical research involving human subjects cannot legitimately be carried out unless the importance of the objective is in proportion to the inherent risk to the subject.

5. Every biomedical research project involving human subjects should be preceded by careful assessment of predictable risks in comparison with foreseeable benefits to the subject or to others. Concern for the interests of the subject must always prevail over the interests of science and society.

6. The right of the research subject to safeguard his or her integrity must always be respected. Every precaution should be taken to respect the privacy of the subject and to minimize the impact of the study on the subject's physical and mental integrity and on the personality of the subject.

7. Physicians should abstain from engaging in research projects involving human subjects unless they are satisfied that the hazards involved are believed to be predictable. Physicians should cease any investigation if the hazards are found to outweigh the potential benefits.

8. In publication of the results of his or her research, the physician is obliged to preserve the accuracy of the results. Reports of experimentation not in accordance with the principles laid down in this Declaration should not be accepted for publication.

9. In any research on human beings, each potential subject must be adequately informed of the aims, methods, anticipated benefits, and potential hazards of the study,

and the discomfort it may entail. He or she should be informed that he or she is at liberty to abstain from participation in the study and that he or she is free to withdraw his or her consent to participation at any time. The physician should then obtain the subject's freely-given informed consent, preferably in writing.

10. When obtaining informed consent for the research project the physician should be particularly cautious if the subject is in a dependent relationship to him or her or may consent under duress. In that case the informed consent should be obtained by a physician who is not engaged in the investigation and who is completely independent of this official relationship.

11. In case of legal incompetence, informed consent should be obtained from the legal guardian in accordance with national legislation. Where physical or mental incapacity makes it impossible to obtain informed consent, or when the subject is a minor, permission from the responsible relative replaces that of the subject in accordance with national legislation.

    Whenever the minor child is in fact able to give a consent, the minor's consent must be obtained in addition to the consent of the minor's legal guardian.

12. The research protocol should always contain a statement of the ethical considerations involved and should indicate that the principles enunciated in the present Declaration are complied with.

II. Medical Research Combined with Professional Care (Clinical Research)

1. In the treatment of the sick person, the physician must be free to use a new diagnostic and therapeutic measure, if in his or her judgement it offers hope of saving life, reestablishing health or alleviating suffering.

2. The potential benefits, hazards, and discomfort of a new method should be weighed against the advan-

tages of the best current diagnostic and therapeutic methods.

3. In any medical study, every patient—including those of a control group, if any—should be assured of the best proven diagnostic and therapeutic method. This does not exclude the use of inert placebo in studies where no proven diagnostic or therapeutic method exists.

4. The refusal of the patient to participate in a study must never interfere with the physician–patient relationship.

5. If the physician considers it essential not to obtain informed consent, the specific reasons for this proposal should be stated in the experimental protocol for transmission to the independent committee (I, 2).

6. The physician can combine medical research with professional care, the objective being the acquisition of new medical knowledge, only to the extent that medical research is justified by its potential diagnostic or therapeutic value for the patient.

III. Non-Therapeutic Biomedical Research Involving Human Subjects (Non-Clinical Biomedical Research)

1. In the purely scientific application of medical research carried out on a human being, it is the duty of the physician to remain the protector of the life and health of that person on whom biomedical research is being carried out.

2. The subjects should be volunteers—either healthy persons or patients for whom the experimental design is not related to the patient's illness.

3. The investigator or the investigating team should discontinue the research if in his/her or their judgement it may, if continued, be harmful to the individual.

4. In research on man, the interest of science and society should never take precedence over considerations related to the well-being of the subject.

16

Insurance and Liability

Marja G. de Jong
Medidas
Wijk bij Duurstede, The Netherlands

Every clinical investigation performed has legal implications for the manufacturer sponsoring the trial and for the investigators performing the trial; in addition, subjects are not exempt.

The European Good Clinical Practice (GCP) requirements and ISO 14155 (including specific national laws in member states) as well as the U.S. Investigational Device Exemption (IDE) requirements include references to compensation indemnification. Insurance may be required by regulatory authorities and/or by the Ethics Committee (EC) or Institutional Review Board (IRB). A manufacturer is responsible for making the proper provisions of compensatory payments to a subject in the event of injury or death arising from participation in a trial.

An overview of countries and their insurance requirements is provided in Table 16.1[*]. Companies are advised to check with their Notified Bodies (NBs) as well as the national authorities in the countries where they intend to open clinical investigation sites.

---

[*]See chapter 9 for further information about insurance.

Table 16.1. Country Overview of Insurance

Country	Requirement
Austria	Insurance required.
Belgium	Insurance required.
Denmark	The Danish Ministry for Health is considering insurance requirements. Patients are already covered.
Finland	No requirements governing insurance. The Ministry for Health is considering drafting clinical investigation requirements, including insurance requirements.

TIP	Although there are no insurance requirements in Denmark, the country has implemented the liability laws ofthe European Union. It is, therefore, wise to have insurance overage.

NOTE	Limit of coverage as well as duration of coverage is determined by the Ethics Committee. The basis of this decision is the type of device being investigated and risks associated with it.

Country	Requirement
France	Insurance (liability and risk) required. Insurance with a French company mandatory.
Germany	Mandatory insurance requirements— Probandenverschierung. Coverage set for a minimum of 1 million DM for death or incapacity to work.
Greece	Insurance required.
Iceland	Insurance not required, but recommended.
Ireland	Insurance required.
Italy	Insurance required.
Liechtenstein	Insurance required.
Luxembourg	Insurance required.
Netherlands	Insurance required (liability).
Norway	Liability insurance is the responsibility of the institution involved in the trial. No specific insurance requirements defined as yet. Recommended.
Portugal	Insurance required.
Spain	Insurance required. Insurance requirements same as for medicines.

Table 16.1 continued on next page.

Table 16.1 continued from previous page.

Sweden	Insurance required.
Switzerland	No insurance requirements at federal level. May be regulated by the Cantons.
United Kingdom	No statutory insurance requirements. May be set by the Ethics Committee. Recommended.
North America (Canada, U.S.)	Insurance coverage required.

NOTE: A universal blanket coverage, entered into with any insurer world-wide, is sufficient for most countries. France is the exception. The key elements in the insurance are liability and risk insurance; an all risks insurance policy is advisable. The coverage and time limit of coverage are crucial elements.

Several factors require consideration when reviewing insurance coverage for clinical trials.

SAFETY

Most devices, prior to their use in a clinical trial, need to be safe. This is of critical importance. For instance, it is negligent for a manufacturer to put a product into clinical investigation which has not been completely tested and technically assessed for risks. Risks are first and foremost eliminated, reduced, and/or warned against in the labelling associated with the device.

TIP	In the United States, the Flood and Drug Administration (FDA) will review the IDE application in great detail to ensure that there are no undue risks associated with the product that may adversely affect the health of the population.

Clinical investigations are not an exact science; anything can happen, and most likely will happen. Adequate insurance coverage is protection in case anything *does* indeed happen.

The duration and coverage of the insurance is crucial. Claims must be covered that may be forthcoming from the use of a device that has not been approved for commercialisation.

| TIP | For all trials, blanket, world-wide coverage is sufficient, provided there is adequate coverage for every country in which the investigation is being planned. Selecting an international insurer who is willing to underwrite the policy is, therefore, critical to the process. Obtain information from the EC or IRB about how it wishes to be informed about the insurance policy and if there are any particular regulations. |

INFORMED CONSENT

The content of the informed consent and the information provided by the manufacturer to the investigator and by the investigator to the patient is crucial. Consent can only be given when concerns have been alleviated. Sufficient information so a subject can make a reasonable judgment is, therefore, necessary. Adequate warnings on the labelling, on the product, and in the instructions for use must be given regarding any risks that have been identified in advance by the manufacturer. Sufficient information regarding these risks needs to be provided to the investigator by the sponsor, and to the patient from the investigator. Failure to do so could give rise to malpractice suits against the investigator.

| TIP | It is not possible to expect to have discovered every possible risk of the device prior to the commencement of the clinical investigation. |

Manufacturers and investigators alike are liable for damage caused by risks that ought to have been discovered and remedied by prior research and testing—but that were not. Remedy of such risks can be through redesign to eliminate or reduce the risk or by warning about them.

TYPES OF INSURANCE

General liability insurance could be considered to be insufficient, although it is the most common form of insurance coverage into by trial sponsors. Risks that may arise from the trial and which, as stated previously, are unforeseen and unreasonable, are to be covered under the insurance as well. This risk liability is related to the fact that a manufacturer is planning to "experiment" with a new device, of which not all the risks are known or can be foreseen at the time of the "experimentation".

TIP	Understand the sanctions. It is unwise to assume that what is applicable on one continent applies to another continent! For instance, in Europe, each Member State in the European Union may regulate legal sanctions differently than in the United States.

In some countries, there are clear responsibilities placed firmly on the manufacturer for harmful side effects arising from the trial. Inherent in this set of laws is a risk development liability coverage. For example, many times, particularly in cases of implants, it is not possible to determine unforeseeable risks until decades after the implantation of the device. Currently, only France and Germany require risk liability insurance as part of the insurance coverage for clinical trials. But this is changing as Europe harmonises its clinical trial processes and the accompanying insurance requirements.

TIP	The time limit for the insurance coverage may be set by the Ethics Committee. The more politically or safety-contentious a product, the higher the coverage that could be required.

The manufacturer is ultimately responsible for compensation. Not every injury may be compensated; hence, risk liability in addition to general liability is important. The "development" risk defence is not always included in liability laws. Many times, it is this aspect that is the most difficult to obtain and may not willingly underwritten by insurers.

Section IV

How to Monitor a Clinical Trial

17

Selection and Training of Clinical Trial Monitors

Willem Ezerman

Parexel/MIRAI
Amsterdam, The Netherlands

Herman Pieterse

Profess® Medical Consultancy BV
Heerhugowaard, The Netherlands

The monitor appointed by the sponsor has two crucial functions. First, the monitor is the liaison between the sponsor and the clinical investigator. Second, the monitor needs to ensure that the study protocol or clinical investigation plan is maintained and that any deviation is reported to and agreed on with the sponsor. One of the main control tasks that the monitor has to perform is the verification that case report form (CRF) is complete, accurate, and consistent in its adherence with the study protocol. Moreover, the monitor has to check the reliability of the entered data by performing source document verifications per a predefined and documented protocol. Before the study is initiated, the monitor has to inform the sponsor about the suitability of the clinical investigator and the facilities. In general, the monitor will act as an assistant, controller, and partner for the clinical investigator.

It is essential that a clinical monitor receive appropriate training in order to perform good quality trials according to the standards of Good Clinical Practice (GCP).

REGULATORY REQUIREMENTS

The GCP guidelines require that

> the monitor is responsible to the sponsor or Contract Research Organization (CRO) for the monitoring and reporting on the progress of the trial and for verification of data. The monitor furthermore must have qualifications and experience to enable a knowledgeable supervision of the particular trial. Trained technical assistants should help the monitor in collection of documentation and subsequent processing. Furthermore, the sponsor has to ensure the ongoing and suitable training of, monitors and clinical research support personnel occurs.

However, GCP does not specify what "qualification" consists of, nor what "experience" a monitor should have in order to function properly, nor what the "ongoing training" of monitors should consist, and what kind of training is suitable and appropriate.

According to the U.S. Food and Drug Administration (FDA), a sponsor may designate one or more *appropriately trained* and *qualified* individuals to monitor the progress of a clinical investigation. The FDA regards the following individuals as qualified:

- physicians;

- veterinarians;

- clinical research associates;

- paramedical personnel;

- nurses; and

- engineers.

Their suitability as monitors depends on the type of product involved. According to the FDA,

> A monitor need not be a person qualified to disclose and treat the disease or other condition for which the test article is under investigation. (FDA monitoring guidelines)

The FDA does not give details on what *appropriately trained* and *qualified* means, and thus leaves it open to interpretation by the sponsor of the clinical investigation.

Standard EN 540 as well as ISO 14155 define the function of the monitor as follows:

> A qualified person appointed by the Sponsor responsible for ensuring the investigator's compliance with the clinical investigation plan and for reporting on the progress of the clinical investigation.

Monitors should therefore be

(a) appointed by the sponsor and responsible to the sponsor;

(b) appropriately trained, and have the scientific and/or clinical knowledge needed to monitor the trial adequately. A monitor's qualifications should be documented; and

(c) thoroughly familiar with the investigational device, the study protocol, written informed consent and any other written information to be provided to subjects, the sponsor's

> Standard Operating Procedures (SOPs), GCP,
> and the applicable regulatory requirement(s).

These elements are the most important aspects to ensure qualified monitors must be used. Requirement (b) however could be in conflict with the FDA statement that a monitor need not be a person qualified to disclose and treat the disease or other condition for which the test article is under investigation. The European Requirements stipulate that the monitor have clinical knowledge, whereas the FDA does not specifically require this.

> It is the obligation of the sponsor to ensure that trials are adequately monitored, and to define the appropriate extent and nature of monitoring. The extent and nature of monitoring is influenced by such considerations as the objective, purpose, design, complexity, blinding, size, and endpoints of the trial. There must be on-site monitoring, before, during, and after the trial; however in exceptional circumstances the sponsor may determine that central monitoring in conjunction with procedures such as investigators' training and meetings, and extensive written guidance can assure appropriate conduct of the trial in accordance with GCP. Statistically controlled sampling may be an acceptable method for selecting the data to be verified.

The monitor needs to receive study-specific training (covering such topics as the objective, purpose, design, complexity, blinding, size, and endpoints of the trial), as well as management training (investigator training and organising meetings).

> The monitor needs to follow the established written SOPs as well as procedures that are specified by the sponsor.

This entails that the monitor is required to receive SOP training in those SOPs that are needed to successfully undertake monitoring activities.

MONITOR SELECTION

In an open market, the people with the highest level of education, training, and experience will typically be hired. This principle also applies to clinical research. Monitors should be selected for the job on the basis of the following:

- Education.
    - an academic degree in medicine, human movement studies, or medical biology, or any other relevant medical/biological/scientific accreditation;
    - nurses with practical clinical trial experience;
    - proficiency in English and the local language; and
    - knowledge of computers and software applications, (i.e., WordPerfect, Word for Windows, Excel, Access, and E-mail).
- Training,
    - GCP training and/or
    - indication training;
- Experience
    - with indications, and
    - with clinical research; and
- Communication and organisation skills,
    - proven interpersonal communication skills;
    - a good sense of human behaviour in problem situations; and
    - accuracy, good humour, and the ability to work and communicate with professionals.

For example, in Eastern European countries, the selection criteria includes that the monitor needs to be a doctor or have an affiliation with a network of doctors, be proficient in English and the local language, and have a thorough understanding of local laws concerning clinical research.

Table 17.1 defines a sample form that could be used to score a candidate for the monitor function. For the monitor function, management should first decide on the weighting factors in relation to the job requirement criteria. In this form (adapted from a form drafted by Hay Management Consultants in 1986), the weighting factors are divided into mandatory, important, useful, and not applicable job requirement criteria. Once the criteria and the weighting factors for the job have been decided, the candidate can be scored during or after the interview. The interviewers can determine to what extent the candidate meets the job requirements (insufficiently, sufficiently, or too well). The form can be a useful tool for standardisation of job requirements for a number of functions in clinical research.

MONITOR TRAINING

Monitors are requred to be trained from the moment they are hired.

Training Programme

In the first six weeks, the new monitor (on probation) should be given a training programme aimed at familiarising him/her with the company and the clinical trial(s) that he/she will monitor. The training programme should consist of both theoretical and on-the-job training.

The execution of this programme is conducted by the study team to whom the monitor is assigned and is closely monitored by an independent mentor. It consists of GCP training, SOP training, company-oriented training, co-monitoring, general and study-specific trial management, and feedback/progress meetings with the mentor. After six weeks, the new monitor is ready to be assessed by the study manager/coach on his/her knowledge of GCP, SOPs, the study(ies) he/she will monitor, and general functioning within the team and the company. On the basis of this assessment, it should be determined whether or not the monitor is ready to proceed.

Table 17.1. How to Score a Candidate

Job Requirement Criteria	Weighting Factor			Results for the Candidate			
	@ @ @	@ @ @	O	Insufficient	Insufficient on Some Points	Corresponds to Job Requirements	Exceeds Job Requirements	Exceeds Job Requirements Considerably
Education							
Academic degree in medicine, human movement studies, medical biology, or any other relevant medical/biological/scientific accreditation							
Nurses with practical clinical trial experience							
Knowledge							
General culture							
Technical knowledge							
Practical knowledge							
Proficiency in languages							

Table 17.1 continued on next page.

Table 17.1 continued from previous page.

Job Requirement Criteria	Weighting Factor				Results for the Candidate			
	@ @ @	@ @ @	O	Insufficient	Insufficient on Some Points	Corresponds to Job Requirements	Exceeds Job Requirements	Exceeds Job Requirements Considerably
Management Potential								
Organising ability								
Personnel selection and assignment								
Motivating ability								
Information/training								
Human Relations								
Communicating ability								
Ability to motivate and convince								
Negotiating ability								
Ability to work within a group								

Table 17.1 continued on next page.

Table 17.1 continued from previous page.

Job Requirement Criteria	Weighting Factor			Results for the Candidate				
	@ @ @	@ @	O	Insufficient	Insufficient on Some Points	Corresponds to Job Requirements	Exceeds Job Requirements	Exceeds Job Requirements Considerably
Creativity								
Analytical mind								
Ability to summarise								
Open mind								
Critical mind								
Spirit or initiative								
Orientation toward Results								
Forecasting ability								
Planning ability								
Decision ability								
Will to succeed								

Table 17.1 continued on next page.

Table 17.1 continued from previous page.

Job Requirement Criteria	Weighting Factor			Results for the Candidate			
			Insufficient	Insufficient on Some Points	Corresponds to Job Requirements	Exceeds Job Requirements	Exceeds Job Requirements Considerably
Cost effectiveness	@	@	O				
Integration of company objectives	@	@	@				

Name: _____

Function: _____

Department: _____

Period: _____

Name 1st interviewer: _____

Function: _____

Name 2nd interviewer: _____

Function: _____

Weighting Factor: _____

@@@   mandatory for this function   @@   important for this function

@   useful for this function   O   not applicable

The activities that the monitor is trained in are defined below:

- co-monitoring with experienced monitors,

  - visiting sites for pre-study visits (PSVs), study initiation visits (SIVs), periodic monitoring visits (MONs), and study close-out visits (COVs);

  - executing monitoring tasks; and

  - evaluating and reporting on visits with experienced monitor and mentor; and

- study management,

  - contact with suppliers in general and investigators in particular;

  - general study management and logistics: how to make a study file, how to execute checking device accountability;

- contact with Ethics Committees or IRBs: how to collect which essential documents, how to assist the investigator with the submission of the project to the Ethics Committee or IRB, and how to arrange for an effective and efficient Ethics Committee or IRB approval process;

- contact with investigators,

  - selection and recruitment of investigators (PSV);

  - site initiation; and

  - site closure;

- in-house trial management,

  - CRF management and processing;

  - query management;

  - secondary or in-house monitoring;

- team meetings; and

- on-site trial management,

  - checking conduct of the investigator, and the quality and progress of the study;

  - checking CRFs: which entries should be verified for consistency and adherence to the protocol;

  - source document verification: which CRF entries should be verified in the source documents for reliability;

  - checking device accountability;

  - checking and updating investigator site file; and

  - meeting with the investigator and site personnel.

Standard Operating Procedures

Training in SOPs should be done in two ways.

1. *Annual SOP update training sessions.* These are organised by the department of Training and Development and supervised by the Quality Assurance department.

2. Study-specific SOP training sessions. These are planned at the start of each new trial. During these sessions, the SOPs are reviewed by the entire study team (consisting of a project manager, data-entry personnel, a statistician, a project assistant, and monitors) for their suitability for this particular trial. The writing of an Operations Manual is initiated to cover for any deviations from the SOPs. The deviations will be documented fully in this Operations Manual in order to keep the research process transparent for future inspection by auditors or inspectors.

Good Clinical Practice Training

Training in GCP may be done in two ways.

1. During the training, the monitor receives the training in the first six weeks, as specified previously. This is general GCP training, consisting of knowledge of the regulatory environment and GCP guidelines (focusing on monitor and investigator responsibilities and archiving).

2. On an ongoing basis, GCP training is organised for the new employee.

General Training/Education

Everyone within a company should be encouraged to share his/her experiences with others. When they have interesting items to present concerning GCP or any other item that may be in the general interest of the company (i.e., general medical issues, new developments in data management), or have been to a conference, the subject of which concerns other people within the company, they should be given the opportunity to present their experience during *refresh* meetings—regular meetings held for this purpose. Such meetings have as objectives to allow feedback of expertise and knowledge into the organisation. It is essential that each employee not only learns from the knowledge but also from the mistakes of others.

18

How to Prepare the Clinic

Herman Pieterse

Profess® Medical Consultancy BV
Heerhugowaard, The Netherlands

Patrick W. Serruys

Erasmus University
Rotterdam, The Netherlands

An investigational site must be adequately prepared in order to meet Good Clinical Practice (GCP) requirements. The sponsor is responsible for selecting appropriately qualified investigators who are trained and experienced in the field of the application of the device under investigation. Moreover, the sponsor should ensure that the investigator and his/her staff are familiar with the background to, and the requirements of, a clinical investigation.

The responsibilities of a clinical investigator include being well acquainted with the clinical investigation plan, or study protocol. Above all, the investigator is to ensure that his/her team will be available to conduct and complete the investigation.

Other responsibilities of the investigation team with respect to the preparation of a clinical study are to

- ensure that any other concurrent clinical investigation does not conflict or interfere with the clinical investigation in question;

- ensure the personal safety and well-being of the subjects participating in the study;

- make the necessary arrangements, including emergency treatment, to ensure the proper conduct of the study;

- ensure an adequate recruitment rate of subjects;

- submit the study protocol to the Institutional Review Board (IRB) or Ethics Committee (EC);

- edit the proposal for patient information and informed consent for the patients; and

- ensure that adequate information is given to the subject (or guardian or legal representative), both in oral and written form, regarding the nature of the clinical investigation.

Should the investigator be asked to function as the principal investigator for the study, the sponsor may assign the principal investigator specific functions, such as acting as chairperson of certain study committees (e.g., safety committee, study events committee, etc.) or presenting the study results at symposia and congresses. The preparation of a clinical investigation in a clinic is a complex process and consists of many steps, as shown in Fig. 18.1. What follows is a step-by-step discussion of this process.

Figure 18.1. Flowchart of the Preparation of an Investigational Site

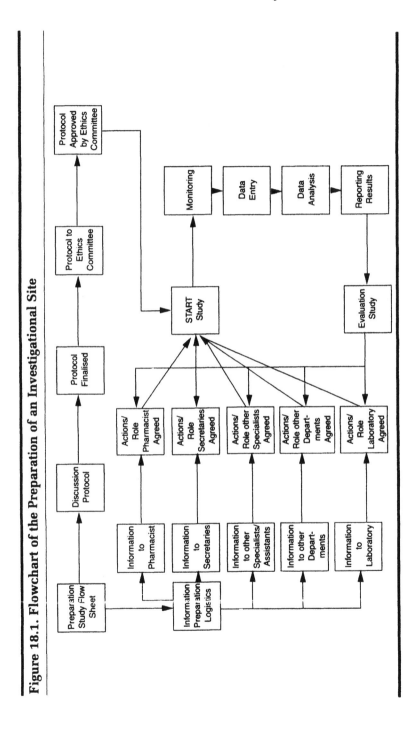

THE PROTOCOL OR
CLINICAL INVESTIGATION PLAN

Once the study monitor has asked the investigator to participate in the study and has explained the outline of the clinical investigation plan, the investigator discusses the study with his/her team. The outline of the first draft of a study protocol should consist of

- the objective of the study;

- the intended patient population;

- the design of the study;

- the actual status of development of the product;

- present clinical experience with the product;

- key efficacy and safety parameters to be measured; and

- methods and procedures to be used during the study.

The outline should be a one-page summary of the aforementioned issues and preferably should contain a time/measurement schedule to visualise what should be measured in the patient population, when, and how frequently.

The investigator is to discuss the outline within his/her department and/or investigation team. The investigator should realise that the following questions have to be answered before a commitment to the study can be made.

- Do we have the expertise and know-how to conduct the study?

- Do we have the facilities (laboratories)?

- Do we have the capacity in terms of personnel?

- Do we have enough patients who might be selected for the study in order to ensure an adequate recruitment rate?

It is essential to investigate the feasibility of a clinical investigation prior to its start. Moreover, it should be questioned if it is medically and ethically justified to start an investigation should it be known beforehand that the conditions regarding feasibility cannot be met. The most important questions are: Can the investigation, as described in the study protocol, be performed in our site? Is it feasible?

With respect to the organisation and logistics within the hospital setting: 'Which other departments/persons (pharmacy, laboratory, nursing staff, secretaries) need to be involved in the study? Are these departments/persons willing to participate? Is it possible to tune the activities of these persons in an optimal way?'

One should also assess the time the investigator, the staff, and others have available to execute the study. It should also be investigated if all of the expertise necessary to conduct the study is present and if all necessary facilities are available. One should remember that many administrative tasks have to be fulfilled.

PATIENT RECRUITMENT

The number of available subjects who comply with all inclusion and exclusion criteria should be screened. Too often, investigators are too optimistic with respect to the recruitment of patients for a clinical investigation. Firm selection criteria limit the number of patients. Moreover, only a fraction of the eligible subjects will give their consent to participate.

The number of available subjects can be calculated in the following way. For retrospective studies, the information is available from previous and comparable studies, or one can make a checklist of the selection criteria and screen the centre(s)'s patient administration. For prospective studies, all new subjects are screened through the use of a checklist during a certain period and an extrapolation is made for the total study period.

Situation I

The recruitment period (A) is the period of absence (B) represents the total duration of absence (holidays, congresses) in the specified recruitment period. Thus, the effective recruitment period (C) is A − B.

The number of subjects who comply with the selection criteria and who visit the hospital every month on average is D. The source for this estimation should be indicated and recorded.

Taking into account that the subjects have to meet the inclusion and exclusion criteria and have to give their consent, a certain percentage of the number of eligible subjects can and will participate in the study. For most studies, between 25 and 50 percent are potential subjects.

$$\_\_\_ \% \times D = E \text{ (the number of eligible subjects per month)}$$

The estimate of the actual number of patients that can be selected for the study (F) is as follows:

$$F = C \times E$$

Situation II

The number of subjects to be included (A) is fixed. The number of subjects with the indication who visit the clinic per month, on average, is B. The source of these estimates should be indicated and recorded.

Taking into account that the subjects have to meet the inclusion and exclusion criteria and have to give their consent, a certain percentage of the number of eligible patients can and will participate in the study. For most studies, between 25 and 50 percent are potential subjects.

$$\_\_\_ \% \times B = C \text{ (the number of eligible subjects per month)}$$

The required recruitment period (in months) (D) is:

$$D = A/C$$

The duration of the treatment according to the protocol is E months. Thus, the duration of the investigation (in months) (F) is:

$$F = D + E$$

When the total period of absence due to holidays, congresses, and so on is G months, then the total duration of the study (H) can be calculated as follows:

$$H = F + G$$

When the start of the study has been planned, then the completion date for the study can thus be calculated.

PRE-STUDY MEETING

Once the questions given in the first section of this chapter have been answered, and the investigational team shows an enthusiastic interest to participate, the investigator needs to communicate this to the study monitor. The study monitor should then arrange a pre-study meeting.

During this meeting, the clinical investigation plan, the conduct of the study, and any other relevant logistical issues should be discussed. Whereas the investigator assesses the knowledge, skills, and expertise of the study monitor, the study monitor assesses the suitability of the investigator, his/her staff, and the facilities (e.g., the laboratories and the pharmacy). An example of a checklist that may be used by the monitor to assess the suitability of the investigator, staff, facilities, and other parties is given in Appendix A of this chapter.

The investigator should analyse the project and ask the following questions of the study monitor:

- What procedure should be followed to obtain approval for the study from the Ethics Committee or IRB? When will all of the documents necessary for submission to the Ethics Committee or IRB be available?

- What is the timetable for the start of the study (first patient in the study), and the accrual time (last patient released from the study)?

- What meetings will be planned in the preparation phase?

- What is the compensation/fee for the time to be invested in the study?

- What extra costs have to be made for the laboratory and the pharmacy?

The following tips will help the investigator avoid delays in the process of obtaining approval from the Ethics Committee.

TIP

Inform the study monitor how the Ethics Committee or IRB works.

- Who is the contact person?

- When does the Ethics Committee or IRB meet in the coming period?

- What is the deadline for submission?

- What documents does the Ethics Committee or IRB wish to receive and how many copies?

- How is the Ethics Committee or IRB to be informed during the of the trial execution and after the trial has been completed?

- Check if the Ethics Committee or IRB is properly constituted.

- Is there a standard submission form?

- What is the administration fee (if any)?

TIP

Find out if other approvals are necessary for the site (i.e., regarding management of the hospital, finances, or scientific committees).

TIP

Find out whether there is a protocol waiting list from the secretariate of the committee.

TIP

Find out whether there are any special indemnification requirements.

TIP

Find out how soon after the Ethics Committee's or IRB's meeting a response is normally issued.

Once the protocol has been discussed in detail and the sponsor has finalised the study protocol, then this document, together with the Investigator's Brochure, the patient information and informed consent, information on the indemnification of subjects in case of injury, and proof of an insurance policy will be submitted to the Ethics Committee or IRB. The aforementioned tips should be taken into account.

INFORMATION TO OTHER PARTIES IN THE CLINIC

When the investigator and his/her team have given their consent to participate in the study, a member of the study team has to inform other departments, secretaries, the nursing department, the pharmacist, and other specialists and assistants who will be liaisons with the investigator and the laboratory. This person could be the investigator or a person specifically assigned to co-ordinate the study in the clinic, (i.e., a study co-ordinator).

A meeting needs to be arranged with those disciplines in the hospital who will contribute to the success of the study. The nursing staff is a discipline that is often forgotten, but which should not be forgotten, because a motivated team of nurses is a prerequisite for the success of the study. In most cases, the study monitor will accompany the study co-ordinator during this meeting.

The study co-ordinator does not necessarily have to be a medically trained person. As long as the investigational team has assigned the medical tasks to a physician, and this physician also takes the responsibility for the entry of the study findings on the case report form (CRF), the organisational and co-ordinating tasks could be assigned to a member of the investigational team who has other basic education and training, but who possesses excellent organisational and communicative skills.

As a first step, the study co-ordinator must prepare a so-called "study flow sheet" based on the measurements and procedures that have to be followed for the study protocol. This could be a flowchart/checklist with information on what has to be done, when, how, and by whom. This "site-specific study flow sheet"

TIP	It is essential to make a "study flow sheet" specific for each discipline.

becomes the basis for discussion with other disciplines involved in the study.

A thorough knowledge of the study protocol and the procedures to be complied with during the study are essential for an effective and efficient study co-ordinator. The better prepared the study co-ordinator is, the more success he/she will have in convincing the other disciplines to commit to participating in the study.

The objective of discussions with other disciplines is to agree on their role and tasks for the study. Should any objections arise, the study co-ordinator should immediately inform the investigator and appropriate actions should be taken to avoid an irreversible situation. Good and effective communication is essential at this stage in the preparation of a clinical investigation.

The following questions could be anticipated to arise sometime during the preparation of a clinical study.

- Are you sure that the investigator has sufficient time to study the protocol thoroughly?

- Does the investigator have ample time to inform the other disciplines within the clinic properly or should someone else do this job?

- At which stage should other disciplines and other personnel be informed about the planned study? Should this be done at the stage that the protocol is finalised and is submitted to the Ethics Committee or IRB, or sooner?

TIP	Information and communication can never be passed along soon enough! When a project is in a preliminary stage, then the information to others should be concise, with the only objective being to communicate properly with the other disciplines.

INSTRUCTION AND TRAINING
DURING THE PREPARATION

During the preparation, the sponsor will arrange meetings to instruct and train the investigator and his/her staff with respect to the procedures that have to be followed for the study. If the study protocol requires specific expertise, then the sponsor might arrange dry runs or test procedures to improve standardisation. In angiography procedures, to evaluate the efficacy and safety of, for example, cardiovascular stents, it is essential that each investigational site performs the angiography procedure in the exact same way. The study site should perform a number of angiography films according to the study protocol; these will be assessed by experts. The experts will qualify the investigational site based on these test procedures.

Other subjects of training and instruction prior to the start of a clinical investigation are

- a detailed discussion of the study protocol and other study-specific procedures;

- monitoring procedures, including device accountability procedures, documentation, filing, and archiving;

- reports of serious adverse events (patient- or device-related);

- documentation of the study findings in the CRF and how to make corrections if any if mistakes are made; and

- information on the product and how to conduct a clinical investigation in compliance GCP.

INITIATION VISIT

The final instruction and training session is the "initiation visit". During this initiation visit, the study monitor, and the study co-ordinator of the site, assess whether all pre-study requirements have been met, and if all disciplines have been properly trained and instructed. Procedures on how to document the study findings on the CRF and how to archive study documents and correspondence also should have been explained. For most studies today, the study monitor will make an investigator file, where all documents will be kept during the course of the study. An example of an investigator file is shown in Appendix 2 of this chapter.

The study initiation visits guarantee "site-specific" training. However, this individual approach has the disadvantage that representatives of other study sites are not present and discrepancies cannot be dealt with immediately. Other sites might have a different opinion about a certain problem. Thus, sponsors also organise "start-up meetings". At these meetings, all investigators and staffs involved in a multi-centre study are invited. The disadvantage of such a meeting is that many people will participate and a personal approach cannot be given. This personal approach, together with the ample time provided to anticipate any problems, is the true advantage of an "initiation visit" to each site.

The initiation visit or the start-up meeting is organised once the study protocol has been approved by the Ethics Committee or IRB. The study devices and other supplies necessary for the study will be delivered or shipped to the investigator or the pharmacist. Then the study can start.

APPENDIX A. PRE-STUDY CHECKLIST TO BE USED BY MONITOR

Suitability

1. Assess the scientific background and area of expertise of the investigator. Obtain a current curriculum vitae that has been signed and dated.

2. List or discuss the investigator's clinical trial experience with previous trials and the interest in the present trial.

3. List all personnel that will be involved in the trial (i.e., co-investigators, research nurses, pharmacists, etc.).

4. Discuss and check if an adequate number of patients will be available. Request to see a subject screening log listing the patients who might be eligible for the present study. If the numbers seem adequate, then record the expected number of patients for the trial.

5. Discuss the timetable for the different phases of the trial, including the expected start of the study and the date when the last patient should have completed the trial.

6. Discuss if the investigator is planning prolonged absences. If yes, then discuss who will replace him/her during these absences.

7. Ask whether the investigator is presently conducting (other) studies.

The Investigator

Discuss the following subjects with the investigator.

8. The investigational status of the study drug.

9. The protocol and study design.

10. The Investigator's Brochure.

11. The CRF.

12. Any suggestions and/or recommendations for the protocol and CRF.

13. Legal responsibilities.

14. Record keeping/archiving of subject and study documents.

15. Insurance of the patients.

16. The necessity to obtain Ethics Committee or IRB approval prior to start of the trial.

17. The obligation to affiliate with or constitute an Ethics Committee or IRB to obtain approval if an Ethics Committee or IRB is not currently available.

18. The necessity to obtain informed consent from each patient before enrollment into the trial.

19. The possibility of an audit by the regulatory authorities or the sponsor.

20. The laboratory procedures.

21. The storage and distribution of the study device.

22. The investigator file.

23. The GCP requirements.

Monitoring Procedures

Have you explained the following items related to monitoring procedures?

24. The procedures for monitoring visits?

25. The device accountability procedure?

26. (Serious) adverse events reporting and handling?

27. The possibility of comparing the CRFs with the source documents?

28. The necessity to obtain accurate and legible CRFs?

Facilities

Discuss the following as related to the facilities.

29. Who will be the responsible pharmacist for the study device if applicable by law? Did you meet the pharmacist and explain the study?

30. Did you visit the pharmacy and check for adequate facilities to store the study device?

31. If there is no pharmacy, are there adequate facilities to store the study device? Who is responsible for ensuring appropriate handling of the study device?

32. Is the laboratory central or a local laboratory?

33. Who will be responsible in this laboratory for the trial?

34. Did you meet the responsible person and explain the (needs of the) trial?

    - Record keeping?
    - Labelling of samples?
    - Reporting of results?
    - Archiving system?

35. Does the protocol require special equipment?

Financial Contract

Discuss the following as related to finances.

36. Financial matters.

37. The contract with the investigator.

APPENDIX B. CONTENTS OF THE INVESTIGATOR FILE

Patient Documents

- Signed informed consents
- Patient identification code list

Study Protocol

- Protocol amendments with a signature and the date when implemented

Study Documents

- Investigator's contract with information on tasks and responsibilities
- Financial contract (this could be an integral part of the investigator's contract)
- The curricula vitae of the investigator, the co-investigator, and other members of the study team (which document that these persons are qualified to perform the study)
- List of the personnel involved in the study with their signatures and initials
- Modification of the list of members of the study team
- Adjustments to the investigator's contract
- Audit certificate stating that an appropriate audit has taken place

Approvals

- Notification to the regulatory authorities, if required

- Approval from the Ethics Committee for the study protocol and for major amendments of the protocol

- The approved final patient information and informed consent

Investigator's Brochure

- Up-to-date scientific product information for the investigator

Laboratory Documents

- List of reference values needed to interpret any laboratory values

- Certificate of accreditation of the laboratory to demonstrate the quality of the laboratory

- Changes of the reference values during study execution

The investigator file should be archived in accordance with national medical/clinical practices laws. Under the laws the patient identification code list and the investigator file could have to be retained for 15 years. The sponsor should be informed where the investigator file is stored and archived.

19

Monitoring Visits

Allison Oliva

Cardiovascular Dynamics
Irvine, CA

Peter Duijst

Aortech Europe Ltd.
Almere, The Netherlands

In clinical research, the study monitor functions as the primary liaison between the sponsoring company or institution and the clinical investigator(s). As such, a monitor is responsible for maintaining a clear communication channel and accurate data transfer between these two parties. Visits made on a regular basis by the monitor to the study sites help to achieve this goal.

The most essential and all-encompassing of the study monitor's responsibilities is the tracking of the clinical trial process. If the study is in its initial phases, the monitor should have a solid understanding of the study protocol and knowledge of the current literature and of specific articles written by the investigator. If the study is ongoing, the monitor should refer to previous visit reports and any correspondence or audit lists to address the remaining issues, and to determine the tasks to be accomplished during the current visit. Monitoring duties will differ depending on the type of visit being conducted (see the section titled 'Types of Monitoring Visits' later in this chapter).

WHAT IS A MONITORING VISIT?

The term *monitoring visit* basically refers to the designated monitor physically visiting the study site. During the visit, the monitor may meet with the principal investigator, the study coordinator, and other study or department staff as necessary.

In general, the term *monitoring* means an action, such as checking, watching, or keeping track of something. However, in clinical research, the term is used in a much broader sense. Clinical monitoring is pro-active in nature and should be utilised to anticipate, early in the study phases, potential difficulties in enrollment rates, data collection, and protocol compliance.

WHAT IS THE PURPOSE
OF A MONITORING VISIT?

The monitoring visit is an important tool, if not *the* crucial tool, of a clinical trial. Frequent visits are key to a well-run trial for a number of reasons. A visit demonstrates sponsor oversight of the project; but even more important, a visit enables the monitor to maintain frequent personal contact with the study staff, which is necessary to audit data during the collection process and to track study material disposition. In addition, face-to-face meetings can often resolve misunderstandings that may hinder study progress.

NOTE | The monitor is the actual "face" of the company at the study site, usually one of the first representatives to contact them, and usually staying in contact over a considerable period of time. Establishing and continuing good relations is not only important for the data collection process, but also for the future relationship between the site and the company once the study has been completed.

WHY ARE MONITORING VISITS NECESSARY?

Currently, the U.S. Code of Federal Regulations[1] (CFR) and the European standard EN 540[2] have no specific requirement for a physical visit to the prospective study site. However, in the Food and Drug Administration (FDA) Monitoring Guidelines[3] and implied in the EC GCP, it is recommended that for a study of medical devices, a monitor must "according to a predetermined SOP, visit the *investigator before, during, and after the trial* to assure that all data are correctly and completely reported". In addition, these guidelines require that a monitor "ensure that the trial site has adequate space, facilities, equipment, and staff" and available patients to conduct the study appropriately. So, although data could be collected by electronic transmission, by fax, or by mail, and the monitor could maintain contact via telephone or other electronic means, it would be very difficult, if not impossible, for a monitor to assess the other required aspects without a physical visit to the study site.

The total number and frequency of visits to each site is based on the duration of the study, the number of patients to be enrolled, and the level of study oversight or management available at each site. These and other variables must be determined and reassessed by the monitor throughout the course of the study[4,5].

 In addition, site visits demonstrate dedication to the project and help to build important personal relationships with the individuals responsible for the study at each site. In the end, these efforts can only improve data quality and compliance.

---

1. U.S. Code of Federal Regulations, 21 CFR 812.20, 812.43.

2. EN 540:1993. 'Clinical Investigation of Medical Devices for Human Subjects,' Section 5.5.

3. *Guide to Good Clinical Practice*, Thompson Publishing Group, December 1993, Appendix IV.

4. EN 540, Section 5a.

5. 21 CFR 312.56, 812.46.

TYPES OF MONITORING VISITS

During the course of the clinical trial, there are four primary types of monitoring visits:

1. pre-investigational site visit;

2. study initiation visit;

3. periodic visit; and

4. study close-out visit.

Each of these visits will be described below, including a list of activities suggested before, during, and after each visit.

Before the Monitoring Visit

Prior to any monitoring visits, the sponsor must identify and select potential investigators. This portion of the clinical trial plan may occur with input from the monitor or other company staff. It may have been previously determined by the scientific advisory board, specialists in the field, or through a search of study sites who perform large numbers of the type of case to be studied.

While the potential investigator can be initially contacted through an introductory letter or telephone call in order to get a general sense of the investigator's level of interest in conducting a study, the monitor may decide to meet in person with the investigator during an initial contact visit.

Scheduling a Monitoring Visit

Before any visit, the monitor should contact the investigator and/or study co-ordinator to verify availability. A monitor should clearly state the purpose of the visit and present a proposed agenda and visit timetable.

For example, for an initial contact visit, the monitor may provide a draft of the investigational plan for review in a 30-minute to 1-hour meeting. In later phases of the study, a monitor may need to conduct a study file audit that requires 2 to 3 days of input from the study coordinator, but only a brief interview with the principal investigator at the end of the visit.

NOTE	The principal investigator is often unavailable for extended meetings but may be able to schedule a meeting between surgeries and other appointments. In many cases, however, the study coordinator will be the prime contact.

At any site visit, a monitor should always provide a forum for the investigator and staff to discuss concerns and impressions about the overall study plan.

Pre-Investigational Site Visit

Before finalising the selection of an investigator, the monitor needs to address several issues:

- Does the investigator have any relevant experience in performing clinical studies?

- How many procedures are performed at the facility on an annual basis?

- How many procedures have been performed by the potential investigator?

- Does the study site have well-equipped facilities, staff support, and an adequate potential patient recruitment pool to complete the study as outlined in the protocol?

Although these factors can be discussed with the investigator in advance, they may not be determined until the monitor performs the pre-investigational site visit. The purpose of the pre-investigational site visit is to verify the ability of an investigator to conduct the proposed study. During this visit, the monitor is responsible for evaluating through research, interview, and observation the education, scientific training, and previous clinical trial experience of the investigator and study staff, as well as the adequacy of study facilities, equipment, availability, and management capabilities of the study site.

NOTE | The terms *site initiation, pre-study,* or *placement visit* are synonyms for *pre-investigational site visit.*

Typically, a pre-investigational site visit will include introductions to the principal investigator, the co-investigators, and study staff; a tour of the facility; review of the study protocol, case report forms (CRFs), patient enrollment, data collection, and Ethics Committee/Institutional Review Board (IRB) approval process; notification of regulatory authorities where appropriate; and discussion of budgetary issues. To facilitate this discussion, a written draft of the investigational plan and other materials should be provided to the investigator prior to the meeting.

If a monitor has not previously met the investigator, an introduction and brief description of the monitor's previous monitoring, and the technical and research experience of the monitor should be provided to the investigator. This description serves to reassure an investigator of the monitor's qualifications and creates confidence in the monitor's capability to oversee and guide the project. The monitor should also be prepared to discuss the investigator's areas of interest. This investigator has been identified and selected because of his/her expertise in a particular area of research. Prior to meeting with a prospective investigator, a literature search for recent articles and related areas of interest should have been performed. This will help to establish common ground and a rapport with the investigator that may be essential for rapid communication later in the trial.

There are many issues that determine the suitability of a study site and ultimately the success of the project. A general list of items to be determined at a pre-investigational site visit is given below.

- Will the investigator perform the study according to the study protocol?

- Does the investigator have realistic expectations about patient enrollment?

- Are the facilities large enough to accommodate patient enrollment?

- Is the patient pool large enough to guarantee reasonable recruitment?

- Is there space to store the study devices adequately and safely?

- Will the study devices be segregated from the regular inventory?

- Who will have access to the devices and how will access be controlled?

- Who will be responsible for device accountability?

- Who will have responsibility for the study data and files?

- Will the current staffing levels and clinic status allow compliance with the protocol?

- Where will the study files be stored and who will have access to the files?

- Are the CRFs clear to the staff and will they be used correctly?

- Will informed consent be required and obtained for all patients?

- Will the necessary source documents be provided?

- What is the process for Ethics Committee or IRB approval?

- Who reports the adverse events and how is study continuation determined?

- Will the study budget need to include compensation for additional lab tests or extra manpower?

 If a potential clinical study site has little or no experience in performing clinical trials, or when innovative complex technology is involved, it may be difficult for the monitor to assess whether the study site and the investigator are appropriate for the study. In these cases, the monitor must ensure that thorough explanations are given regarding the time and cost aspects associated with the study and clarify in detail the responsibilities of the investigator and/or study co-ordinator regarding all aspects of the study, before selecting or eliminating a potential study site.

During the pre-investigational site visit, the monitor must thoroughly review the study design and rationale with the investigator and staff to make a determination of their willingness and ability to conduct the study. If the study protocol is still in draft form, the monitor may invite comments and feedback from the clinician and staff. Protocol or CRF modification at this stage can often eliminate the need for revisions and amendments during later phases of the study. The monitor should explain their role and responsibilities and outline the tasks and areas of responsibility required of the investigator and study site staff. The clinical staff should begin to identify the methods and personnel responsible for access and control of all study materials.

CRFs should receive particular attention at this point. A review of the total volume, and the number of forms needed per patient, will provide the investigator and study staff with a quick and concrete idea of the amount of work required, and the demands that participation in the study will impose on their office or department.

A study guide, containing detailed instructions for completing all CRFs, as well as a schedule indicating form collection time points, should be provided. The study guide can serve as an introduction to the process of a clinical trial and may clarify the resources needed at each phase of the study.

NOTE After reviewing the number of forms required for each patient, an investigator may realise the need for an on-site study co-ordinator, or additional administrative support, which should be considered when discussing the total study budget.

It is recommended that the monitor provide the investigator with a study budget proposal that includes an overall grant and a breakout of specific patient and study site costs. A more complete discussion of budgets and resources can be found in chapter 10.

Final budget decisions do not need to be made at the pre-investigational visit. Any requests for additional monies or changes to the budget proposal should be carefully noted, but the monitor does not need to make any promises or statements other than to assure the clinician that a response will be provided at a later date. The monitor and investigator can identify particular requests that are considered necessities, as opposed to wants.

The payment schedule and method of payment should also be discussed at this time. Grant payments may be made at specific time periods during the study or may be linked to milestone achievements of patient enrollment, study form completion, and so on. An example of a regular payment schedule is shown in Table 19.1.

Budgets and grant requirements will vary widely depending on the size of the study, type of device, experience of the investigator, and size of study centre and staff, as well as specific institution grant requirements. The method of payment and allowable level of compensation for clinical studies may also differ at each institution. The monitor should verify that the investigator is complying with all institute or government regulations before finalising a study grant.

Since the clinical study approval process and informed consent requirements also vary with each hospital and institution, the monitor needs to identify what type of review committee exists, how the committee is formed, the chair and other members of the committee, and the members' qualifications. Whenever possible, the monitor should request a copy of the written policies and procedures governing the Ethics Committee or IRB review process and schedule of approval timelines at that same institution.

The monitor should also ascertain who will be responsible for contacting the Ethics Committee or IRB and providing the

Table 19.1. Study Budget Proposal

Study Milestone	Percentage of Total Budget	Total Awarded
Study onset	25%	25%
50% patients enrolled	25%	50%
All patients enrolled	25%	75%
Final report produced	25%	100%

necessary information. Often, the investigator will contact the Ethics Committee or IRB and provide them with the study protocols, the Investigator's Brochure, and additional product information. However, in cases where the Ethics Committee or IRB operates regionally or nationally, the sponsor and monitor may be responsible for this task.

NOTE Despite regional or national approval, individual hospitals or research institutions may require additional review by an internal review board.

At the conclusion of the pre-investigational site visit, the monitor should be able to make an accurate assessment of the investigator's interest in the study and whether a sufficient level of understanding exists to execute the study successfully. The goal of the visit is to establish a relationship with the investigator, to be built on in future meetings, and to ensure that all parties involved understand exactly what is expected once the study begins, and as the study progresses through each phase to completion. Results of the pre-investigational site visit should be documented in a written report for discussion with clinical management. This report should be used to determine final study site eligibility, budget requirements, and future monitoring levels.

Pre-Study Training and Documentation

In the time period between finalising study negotiations and the initial use of the investigational device, the monitor is responsible for providing the study staff with the appropriate training materials and collecting all of the necessary regulatory documentation. It is recommended that a group meeting, including the principal investigator, co-investigators, other participating clinicians, nurses, study co-ordinators, laboratory personnel, and relevant administrative staff, be held to review the study protocol, study forms, patient enrollment, and study timetables.

The monitor should go through the protocol and study forms with the study staff *page by page,* checking clarity and conformity, to ensure that the need for compliance with GLP

A group meeting creates a team approach to the study project. All participants should be encouraged to participate freely, and to raise questions from different perspectives that can be addressed prior to use of the device and collection of data from a study patient. In subsequent visits, it is important that the monitor revisit any issues and resolve concerns about the study progress.

and GCP guidelines is thoroughly understood. The monitor should address the issues of patient informed consent, patient inclusion and exclusion criteria, the randomisation scheme (if applicable), and adverse event definitions and reporting procedures.

If training has not already been provided, the monitor should explain the correct procedures used to enter data on each form, the acceptable laboratory units, medical or procedural abbreviations, and the appropriate method for the correction of data and the documentation of any changes. The monitor should emphasise (and reiterate as necessary) the importance of accuracy, completeness, and legibility when entering data on the study forms.

For highly complex devices, or a newly developed procedure, the sponsor should consider specialised training in an animal model or "wet lab" to give the clinical staff "hands on" experience with the device or procedure prior to the use of the investigational device on a study subject. The sponsor may decide to produce a clinical training video that the monitor can use to outline the scope, objectives, and outcome measurements of the study, and to demonstrate use of the investigational device to a group audience.

Before the investigational device can be utilised on a human subject, specific study documentation must be completed and collected by the study sponsor. The necessary documents are the

- signed investigator's agreement;
- curricula vitae of the principal investigator, co-investigators, and research staff;
- study protocol signed by the investigator;

- study approval letter/form from the Ethics Committee or IRB;

- list of Ethics Committee or IRB members and qualifications;

- approved patient informed consent;

- patient information package;

- signed patient insurance policy;

- signed study payment/grant agreement (if applicable);

- laboratory certificates; and

- laboratory normal values.

The monitor is responsible for ensuring that these documents have been collected, and that all members of the study staff have a complete understanding of the study responsibilities, before agreeing to or authorising release of the clinical product to the study site.

The Study Initiation Visit

In some cases, the pre-investigational, training, and study initiation visits are combined into one visit. For example, if the sponsor has had previous experience with the investigator and study site, or is already conducting a study at the same facility, it may not be necessary to make a separate pre-study assessment. However, the protocol, data forms, data collection schedule, and responsibilities for study conduct must always be carefully reviewed with the investigator and staff before actual study initiation.

The study initiation visit takes place at the first use, or one of the first uses, of the investigational devices on a human subject. The sponsor often sends a monitoring team, consisting of the clinical monitor, an engineer or technical services representative, and a medical advisor, to observe the initial use of the device.

The monitor should be available for general and technical support but generally does not interfere with the practice of medicine, *unless not doing so would cause the patient to be placed at risk.* Immediately after the first case, the monitor should debrief the investigator and collect his/her impressions of the use of the device. The monitor should assist the investigator or study co-ordinator in completion of the first set of CRFs.

The monitor should be aware of any discrepancies between the recommended and actual use of the device (this may indicate the need for a protocol revision) and any confusion or questions regarding transfer of data to the CRFs. The monitor should clarify adverse event definitions and the proper procedures for reporting these events, as necessary.

Periodic Visits

On a regular basis, the monitor will visit each of the clinical sites throughout the course of the study. The frequency of periodic visits depends on several factors, but primarily on the study progress and the availability of the investigator or study co-ordinator[6,7].

An experienced, well-trained, and well-organised study site will obviously require less oversight from a monitor than a study site without clinical trial experience or one that is lacking a dedicated study staff. The monitor must constantly assess the level of contact necessary with each site to ensure ongoing interest and compliance of the investigators and staff.

The objective of the periodic visit is to ensure that the study is being conducted according to the protocol and to verify the data collected to date. A general overview of items for review at a periodic visit is given in Table 19.2.

At the periodic monitoring visit, the monitor will check the data entered on CRFs for accuracy and completeness. This is best done by comparing each item on the CRF with the patient's chart and hospital file. In other words, the monitor will "audit" the CRF against a source document to verify that the data exists and has been transferred correctly to the CRF. An example of a file audit procedure can be found in the Appendix of this chapter.

---

6. EN 540, Article 5a.

7. 21 CFR 312.56, 812.46.

Table 19.2. Checklist for Periodic Monitoring Visit

Subject	Items to Check
Protocol compliance	Have protocol deviations occurred?
	If so, were they reported, justified, and documented in the appropriate manner?
	What actions were taken to avoid further deviations?
Adequacy of site facilities, staff, and equipment	Have there been changes to the clinical facility or laboratory?
	Have there been changes in study staff or administration?
	Have the changes impacted study progress?
Patient information and informed consent	Has informed consent been documented for all patients?
Accuracy and completeness of study files	Is the investigator maintaining a study file?
	Have the CRFs been completed legibly and accurately?
	Have all adverse events been reported and documented?
	Is source documentation available for data verification?
Storage and distribution of clinical supplies	Have there been any problems with device shipment or receipt?
	Are the study devices being stored properly and distributed in a controlled manner?
	Is a device inventory available and correct?
Study/laboratory approval	Are all regulatory approvals current?

NOTE Personal visits demonstrate concern about the overall study progress; and cannot be underestimated. Therefore, it is highly recommended that at least one *joint* visit per year be made by the monitor and a member of the sponsor's executive staff. A visit from the management team demonstrates that the sponsor has a vested interest in the welfare of the investigator and interest in maintaining the investigator's involvement in the project.

Monitoring results should be discussed with the investigator and study staff, and specific corrective actions should be determined to avoid repetition of mistakes on future forms. The monitor may need to provide additional training or a review of correct data handling procedures to the study staff.

If time permits, study forms should be corrected during the site visit. All corrections should be discussed and approved by the investigator and study staff. In general, corrections should be made by drawing a single line through the error, writing the correct answer adjacent to the error, and initialing and dating the corrected value. If corrections cannot be made during the visit, the monitor should make a list of all data to be corrected and leave a copy at the site, or send a copy to the site after the visit has been concluded.

NOTE	Erasure or "white out" of data should not be allowed on CRFs.

At the completion of each visit, the monitor should conduct a brief meeting with the investigator and staff to review their findings. If necessary, 'action items' should be assigned, with agreement for correction or resolution prior to the next scheduled visit.

After the Monitoring Visit

At the completion of each site visit, the monitor should prepare a site visit report for the study files, summarising the activities of the visit. A cover page for the report should include the study site name; the investigator's name, address, and telephone number; the monitor assigned to the study; and a list of participants present at the monitoring meetings. It may also include a running tally of the number of patients enrolled in the study, and patients who have been excluded, are deceased, or lost to follow-up in that time period. At each subsequent visit, these numbers can be reviewed and will provide simple accountability of all active patients. A checklist may indicate any changes that have occurred at the facility, the study site, or in the investigator's compliance with the study protocol.

Finally, a summary should include a list of documents or devices collected during the visit and a discussion of any other topics that require follow-up. A copy of the summary may be sent to the investigator as a reminder of agreements made during the visit and tasks to be completed prior to the next monitoring visit. An example of a site visit report can be found in the Appendix of this chapter.

Study Close-Out Visit

Once patient enrollment has been completed, and all CRFs have been collected and final changes and corrections have been made, the study will be terminated and the monitor will make the final study close-out visit. This may also be called the *study termination* or *final close-out* visit.

At the final visit, the monitor must verify that all informed consents were reviewed and signed, that CRFs have been collected and audited, and that study records will be retained as required by law. For U.S. studies, records must be maintained by the investigator for 2 years after a marketing application is filed. If no application is filed, or if an application is denied, records must be retained for 2 years after the date of study termination[8]. In the European Community, for studies with devices, data must be archived for "an appropriate time"[9], which according to EC GCP guidelines should be for a period of at least 15 years after completion or discontinuation of the study[10].

Product inventory must be verified, and any remaining devices must be collected or returned to the sponsor for disposition. The Ethics Committee or IRB must be notified of study completion, and a final report should be provided by the principal investigator to the sponsor and the Ethics Committee or IRB where applicable.

A final report should contain a description of the study methodology and design, data analysis procedures, and a

---

8. 21 CFR 312.57

9. EN 540, Section 5.6.22

10. Good Clinical Practices for Trials on Medicinal Products in the European Community, July 1990, Section 3.17

Table 19.3. Final Report

1. Date of study completion	
2. Total number of patients enrolled in the study	
3. Total number of patients who completed the study	
4. Number of patients who were dropped from the study prior to completion	
5. Type and severity of adverse events Summary of unanticipated, life-threatening, or fatal adverse events Statistical analysis of key data points Critical evaluation and final summary of study results	
6. Summary of unanticipated, life-threatening, or fatal adverse events	
7. Statistical analysis of key data points	
8. Critical evaluation and final summary of study results	

critical evaluation of all study results. A list of essential final report components is shown in Table 19.3.

DATA CHECKS AND FILE AUDITS

When checking the CRFs, the monitor will review the data entered for accuracy and completeness. To verify that the data exist in the patient and hospital files and have been entered correctly, the monitor will audit the CRFs against the source documents.

Generally, the following approach is taken. A set of patient files (10–20 percent of the total patient population) is randomly selected (by statistical methods), and key data points are

Table 19.4. Data Verification Schedule

Verify for All Patients	Verify for 20% of Patients
Patient ID number/initials	Patient medical history
Signed informed consent	Physical examination
Patient eligibility criteria	Laboratory results
Major efficacy/performance data	Non-key efficacy data
Adverse events/complications	Non-key performance data

checked against the source documentation for all patients. Examples of key data points that should be verified for all patients and data that are generally checked for a percentage of patients is shown in Table 19.4.

In the majority of clinical studies, especially in trials with large patient populations, source verification of every data point is impossible. However, if it has been more than 2 years since the initiation of a study and a file audit has never been conducted, or if a random file audit shows that more than 15 percent of the data have been incorrectly entered or is missing, the monitor should consider a 100 percent source verification audit of the study files[11]. One hundred percent verification is a huge, time-consuming process and should not be necessary if the study has been properly monitored since initiation.

THE DO'S AND DON'TS OF CLINICAL MONITORING

To be successful in monitoring a clinical research project, a thorough understanding of the GCP principles is a basic necessity. As important, however, or even more so, will be the attitude of the monitor. When the monitor demonstrates a lack of

---

11. Gerlis, L., and M. E. Allen. 1987. *The Scrip Guide to Good Clinical Practice. Documenting Clinical Trials to U.S. FDA Requirements.* Richmond, Surrey, UK: PJB Publications Ltd. Chapter 4, Section V, point B3.

interest in the study, study sites will gradually lose their interest as well. Obviously, this will be reflected in the data quality. A number of considerations in relation to monitor attitude are given in the following sections.

Do's

Be Prepared

Before meeting the investigator, search and read selected articles. Bring all materials, copies of forms, files to be audited, stationery supplies, and so on to the monitoring visit.

Be Honest

Inform the investigator(s) as completely as possible within the boundaries of your authority. If you are unable to reveal certain information, or you are unsure whether certain information is proprietary or not, refer the matter to your supervisor. Give investigators an approximate timetable for a response to their questions.

Listen/Communicate

Solicit comments from the investigator and study staff. If you are willing to listen, the study staff will inform you of potential problems before they can adversely affect the study or schedule. If you have a non-compliant investigator, give him/her a forum to express his/her dissatisfaction.

NOTE | You do not have to know all of the answers at the monitoring visit. Make careful notes of the major points of contention and take the information back to your supervisor or executive staff for the appropriate response.

Frequent communication is necessary to ensure an accurate and timely study. It is much easier to detect and rapidly correct errors or make protocol revisions when the sponsor and investigational staff communicate on a regular basis.

Negotiate Wisely

One or two topics in a discussion should be flexible. Leave room for negotiation and try for a win-win situation so that all parties are comfortable with the outcome.

Keep Your Promises

Once you have made commitments to an investigational site, be prepared to keep them. If, due to company changes, you are unable to honor your commitment, inform the investigator and staff as soon as possible, and present alternative solutions.

Don'ts

Don't Go Below Your Bottom Line

Accepting the lowest bid has become a well-known business practice. This is NOT how a sponsor should select investigators, and may actually cost more due to protocol and budget revisions within the time period of the study. Negotiations MUST be conducted within the framework of a mutually agreed upon protocol, budget, and timetable. DO NOT allow a study site to force you to accept an initially low budget, nor a reimbursement schedule the sponsor cannot support.

Don't Lose Your Cool

Negotiations require time and patience. If you and the investigator cannot come to an agreement, reschedule the meeting to allow both sides to consider possible options, or to consult with a third party.

Don't Continually Reschedule a Study Initiation Date

Clinical studies require a great deal of preparation by both the sponsor and the investigator. Occasionally, a study may be delayed due to regulatory issues, or lack of sufficient product or study subjects. But, if rescheduling occurs repeatedly, you may find that the investigator is no longer interested or available, and may not be willing to delay patient treatment. In addition, delays may create a lapse in institute approvals or require contract re-negotiation. Poor planning by a sponsor is no excuse for inconveniencing the investigators!

Don't Accept Protocol Non-Compliance nor Allow Incorrect Data to Go Undetected

If you become aware of a protocol violation or uncorrected data, you MUST inform your superiors. Civil and criminal charges have been levied against both companies and individual employees who knowingly accepted false data.

Don't Give Up

Clinical trials of medical devices have become long, involved processes that require lots of money and many years of planning and follow-up. However, it is possible to complete a successful clinical trial and reach the medical marketplace. Remember, your goal as a monitor is to facilitate all phases of the clinical trial, and, in doing so, achieve market clearance of a new device or procedure that will benefit patients worldwide.

APPENDIX. FORM EXAMPLES

Biomedical Company	FILE AUDIT

Study no.: _____ Patient ID No.: _____

Investigator: _____ Co-investigator 1: _____
Center: _____ Co-investigator 2: _____

Forms	Check and Record the Number of Forms Collected Record Major Event Dates								
Consent Form									
CRF1									
CRF2									
CRF3									
CRF4									
CRF5									
CRF6									
CRF7									
CRF8									
CRF9									
CRF10									
Date Verification									

Biomedical Company	**FILE AUDIT NOTES**	

Study no.: _____ Patient ID No.: _____

Investigator: _____
Center: _____

_____
_____
_____
_____
_____
_____
_____
_____
_____
_____
_____
_____
_____
_____
_____
_____

Date Verification.—Record the date that the forms were verified against the patient files at the center.

_____   _____
Signature of Auditor   Date

20

In-House Monitoring by the Clinic

Herman Pieterse

Profess® Medical Consultancy BV
Heerhugowaard, The Netherlands

Elisabeth C. M. van der Velden

Deventer Hospitals
Deventer, The Netherlands

The objective of this chapter is to describe the tasks and activities that have to be performed by the investigational team after the clinical study has started, in order to ensure that the study progresses effectively and efficiently. The chapter concentrates on the investigator's team. The investigator will, in most cases, appoint an in-house study co-ordinator who will perform the majority of the activities described below, in close co-operation with the study monitor.

Before the clinical study starts, it is essential that each member of the investigational team involved in the study has been appropriately trained and instructed (i.e., each member is qualified to do the required tasks). It is the responsibility of the study monitor and the in-house study co-ordinator or research nurse (depending on the distribution of tasks within the clinic) to ensure that everybody knows and does what is expected of him/her.

In practice, a number of logistics problems may arise that seem trivial, but these should not be disregarded. The study co-ordinator and the study monitor should be aware of these problems. Some of these logistics problems are as follows.

- Where should a patient be put for a 30-minute rest period?

- Where in the clinic should a patient wait for the next assessment?

- How do you arrange for the same staff person to perform critical measurements (e.g., ergometry) in any given group of patients (to standardise methodology)?

- How do you plan the visits for the trial during the office hours/consultation hours of the investigator?

- How do you respond to secretaries who think that non-study patients need more attention than study patients?

- What if no beds are available in the hospital on the day you have planned a study?

- Who decides which patient will be treated first when two patients arrive simultaneously at the hospital?

- How do you achieve optimal patient registration and recording and transfer of data between different shifts of nursing staff?

- How do you motivate an assistant physician to get up during the night and take a blood sample?

- Why do simple things so often go wrong, like the execution of measurements at the right time?

- What do you do with a pharmacist who does not want to deliver the devices (e.g., medication) outside normal office hours?

TIP	Make a list of the drawbacks and pitfalls that can be anticipated for the study and discuss these with the study team.

After the study has started, the study co-ordinator has several tasks to accomplish, including the following:

- completing all (or parts) of the case report form (CRF);

- acting as liaison between the different disciplines involved in the study (i.e., laboratory, nursing staff, department personnel, pharmacist);

- solving any trial-related problems in close co-operation with the study monitor;

- assisting the investigator with organisational tasks (e.g., liaison to the Institutional Review Board [IRB] or Ethics Commitee [EC]);

- preparing the monitoring visits for the study monitor.

CRF COMPLETION

The study co-ordinator, the research nurse, or any other assigned staff member will perform tasks with respect to the completion of the CRFs. In some trials, the investigator will complete the CRFs; in other trials, the investigator's staff will perform parts of this task.

In order to avoid any data queries later in the process and to ensure effective and efficient visits by the study monitor, the person who completes the CRFs must perform this task as accurately, consistently, and reliably as possible. To achieve this goal, a number of activities should be performed by whomever completes the CRF (either the investigator or the study co-ordinator). Figure 20.1 is a flowchart of CRF completion.

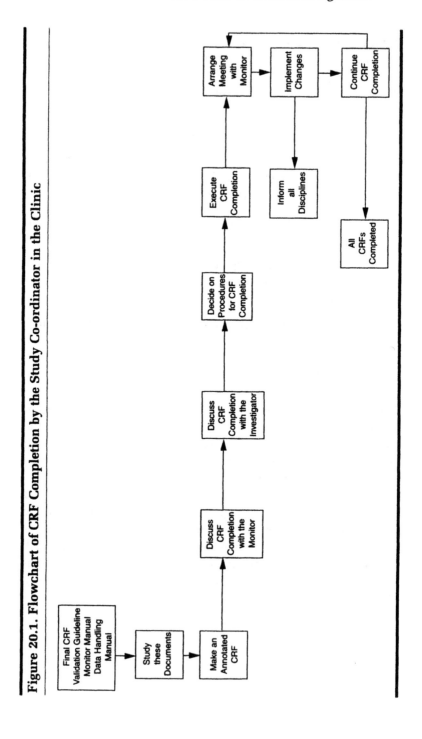

Figure 20.1. Flowchart of CRF Completion by the Study Co-ordinator in the Clinic

Procedure for Effective and Efficient CRF Completion

The Preparation Phase

The person responsible for the preparation phase is the person who will complete the CRF in the clinic. During the preparation phase, he/she should request the following documents from the study monitor:

- final printed CRF;

- CRF validation/completion guidelines;

- monitor manual; and

- data handling manual.

The *CRF validation/completion guidelines* are written guidelines that describe which entries of the CRF will be checked. This includes both the visual checks to be performed by the study monitor and the translation of computer consistency and existence checks into normal language (i.e., which range checks will be performed, which data checks will be performed by the computer-error check, and which consistency checks will be programmed). Consistency checks are cross-checks to test data consistency for related data fields in the CRF.

The *monitor manual* is a manual describing the tasks and responsibilities of the study monitor. In this manual, all of the monitor's tasks are fully described: How is completion of the CRF controlled? Which data must be checked with which source documents? How is device accountability arranged? How is the shipment of the devices arranged? How are documents filed and archived?

A *data handling manual* is a manual made by the database manager. It shows the design of the database; the tasks, responsibilities, and authority of involved data management personnel; and the data entry and data editing procedure. The study co-ordinator or any other assigned person in the clinic should be fully aware which entries will be checked and controlled later on during the process.

If each discipline working on the study—both within the clinic and at the sponsor's office—knows what is to be expected from everyone who works on this project, then extra control activities can be avoided. This means that work does not have to be performed twice. Hence, the cost of achieving a database that completely reflects what has been done in the clinic will decrease.

Study the documents thoroughly and pay special attention to

- the instructions for completing the data fields correctly;

- the instructions for correcting the data fields;

- which data fields should be consistent with which others; and

- which specific correcting instructions to use (e.g., color pen, crossing or ticking fields, etc.).

Fill out an "annotated CRF" with remarks, tips, and "be aware of's" to help guide others through the pages of the CRF. This could be a joint activity of the study co-ordinator and the monitor.

Discuss with the monitor, or obtain from the monitor manual, the sources that will be used for the trial, and which data should be extracted from these sources. Knowing the specific sources will help avoid conflicts of information.

Discuss the following questions with the investigator(s).

- Which data should be documented in the sources and how?

- Who will sign the CRF?

- Which entries will be completed by the investigator? By the study co-ordinator or research nurse?

- Which procedure should be followed for the "secondary review" of the data? (Who will check who, what, and how?)

At the end of the initiation visit organised by the study monitor, ensure that the following items are discussed with respect to the completion of the CRF.

- Which accuracy checks should be performed and how?

- Which consistency checks should be performed and how?

- Which CRF entries should be documented in the source?

- Which documents and other study supplies should be prepared for the monitor visit?

- How will protocol issues, such as drop-out management and protocol violations, be handled?

- When and how does the monitor arrange the monitor visits?

Mid-Study Phase

After the study has started and the first few CRFs have been completed, arrange a meeting with the study monitor and discuss the following.

- Were there any problems in completing the CRFs?

- Are there any discrepancies between actual practice and the guidelines given on the annotated CRF?

- Are there any problems related to protocol adherence?

- Are there any procedural problems with respect to the conduct of the study?

Document all revised agreements that have been made with the monitor. Adjust the annotated CRF based on the aforementioned discussion and inform the other disciplines of any changes. Repeat these steps if any problems arise or any situations occur that were not foreseen. Communicate effectively and efficiently with the monitor.

When this procedure has been followed by the study co-ordinator in co-operation with the study monitor and all agreements have been appropriately documented, then adequate CRF completion can be ensured and queries that would have to be resolved later during the trial can be avoided. The control tasks of the study monitor will become easier once the investigator and his/her staff perform one of their principal tasks (e.g., CRF completion) according to the procedure previously outlined.

TIP	The monitor should ensure that the investigational team has followed the CRF completion procedure.

Investigators and study co-ordinators (e.g., research nurses) are confronted with various kinds of CRFs and adverse event forms that differ from company to company. It is hoped that the industry will standardise the various forms used in clinical studies. Although this is currently a dream, the future will bring more standardised formats for CRFs. For now, the investigator and his/her staff have to deal with the forms as presented to them. It is absolutely essential that the preparation for the CRF completion takes place according to the procedure previously described.

LIAISON WITHIN THE CLINIC TO OTHER DISCIPLINES

Although the study monitor is the primary person responsible for the organisation of the study, the study co-ordinator, who is always present at the investigational site, should and could act as the representative of the monitor. In the preparation phase, the study co-ordinator informs and instructs all the other disciplines involved in the study. This means that a working relationship starts to grow. Thus, the study co-ordinator is the right person to continue these activities once the study has started.

Periodically, the study co-ordinator should communicate with the laboratory personnel, the pharmacist, the nursing staff, and any other discipline in the hospital setting to keep track of the study's progress, to help solve problems, and to listen attentively to those persons. Based on a qualitative and quantitative knowledge of the progress of the study, the study co-ordinator is the first and best person to signal, analyse, and solve problems.

| TIP | Listen to the things that are not said: Non-verbal communication says more on how to act in a certain situation and how to anticipate future problems. |

STUDY CO-ORDINATOR AS ASSISTANT TO THE INVESTIGATOR

The EC/IRB must be notified of any modifications to the study protocol (i.e., amendments). This is the responsibility of the investigator, but the study co-ordinator could act as the investigator's representative. Amendments will be differentiated between minor amendments and major changes. For minor amendments, the Ethics Committee or IRB just needs to be notified, and the study co-ordinator makes sure that the notification has been properly documented. For major amendments, defined as those amendments affecting the safety of the subjects participating to the study, the Ethics Committee or IRB must first issue an approval.

The Ethics Committee or IRB has to be informed of any serious unexpected events that occur and any other events that endanger the safety of the subjects participating in the study or that jeopardise the investigation as such. If the Ethics Committee or IRB wishes to be informed of any trial-related specific issues, then the study co-ordinator should ensure that the Ethics Committee or IRB is informed regularly with respect to those issues.

The study co-ordinator will check that the study is running according to the study plan. In addition, the study co-ordinator will keep track of logistic matters related to the study project plan: the stock of study devices, the delivery of the study devices to patients, and the expiration date of the devices. He/she will give a signal to the study monitor, who will solve any logistic problem for the study site.

Filing and archiving can also be part of the tasks of the study co-ordinator. In the investigator study file, the following documents should be filed during the execution of the study:

- signed and dated informed consents;

- signed and dated protocol amendments, if applicable. (The amendment date indicates the date the amendment became valid—i.e., the date the amendment was approved.);

- changes to the study staff must be documented (curricula vitae should be obtained for any study staff involved in the study);

- any modification to the investigator contract; and

- changes in the normal ranges of laboratory parameters, if applicable for the study.

The study co-ordinator could assist the investigator with the task of informing the patients about the study and requesting their consent to participate. According to the Declaration of Helsinki and Good Clinical Practice (GCP), a patient has to be informed on the benefits and risks of the study and should have ample time to consider his/her participation.

| TIP | Start with the inventory of eligible patients early in the preparation phase of the study. |

The medical records for each patient participating in the study should be kept up-to-date and should contain the information as agreed on with the sponsor (which has been documented in the monitor manual of the study). This medical record should at a minimum contain,

- a remark that the patient has agreed to participate in the respective study;

- the patient identification code (e.g., the randomisation code);

- the start and finish date of the investigation for that patient;

- any serious adverse event (clinical- or device-related), if applicable;

- relevant information concerning the medical course of the patient (e.g., a change in the concurrent medication; reports of investigations performed, such as laboratory results, X ray, etc.) even if not required by the study protocol; and

- premature withdrawal from the study with the reasons why, if applicable.

The patient identification code list should be kept up-to-date. If in an emergency situation other patients should be informed immediately, then the investigator can act through this list.

The study co-ordinator could assist the investigator with the following patient-related organisational and logistic matters:

- making appointments for the next visits; and

- reminding patients of their tasks during the period in between visits (e.g., to complete the patient diary).

The study co-ordinator is the person who should report serious adverse events to the study monitor. Any adverse event that occurs during the performance of a clinical study has to be documented in the CRF (see chapter 23). The fact that the event is or is not related to the device does not matter. Adverse events always have to be followed-up until resolution; the course has to be documented.

Adverse events must be reported to the sponsor immediately upon their occurrence. This can usually be done by telephone, but confirmation by fax should be obligatory. Many companies provide specific forms to report adverse events. Moreover, in the clinical investigation plan, the definition of adverse events needs to be given in detail.

HOW IS QUALITY ACHIEVED IN CRF DATA?

A special problem could arise when clinical trials take place in a Coronary Care Unit (CCU) or any other in-clinic department. Staff changes take place several times per day and the transfer of information from the subjects will not always be done properly. This means that the study co-ordinator should give emphasis to this issue and should discuss the drawbacks and pitfalls with the persons involved. The co-ordinator should motivate and convince the personnel that it is essential that a proper transfer of the findings of the previous shift takes place.

Under all circumstances, the study co-ordinator should communicate to the staff of the CCU that patients not participating in a clinical trial may always be allocated to the available beds. This means that spare beds should be held available for any patient who will not be allocated to the trial. If the study co-ordinator does not obey this rule, then there may be problems with the staff. Because normal, routine patient care is the CCU staff's first priority, it should be remembered that trial patients are often an extra burden to the nursing staff.

The study co-ordinator needs the co-operation of the nursing staff and the assistant physicians because no study co-ordinator can be available day and night for the trial. This means that the study co-ordinator should be sure that the staff is motivated to perform trials. Even more important is the task of keeping the staff motivated.

HOW IS STAFF MOTIVATION MAINTAINED?

The following 10 steps will create and foster a good working relationship between the staff and the investigator/study co-ordinator:

1. Explain the trial and stress the importance of accuracy.

2. Involve the staff in the early preparation stages of the trial.

3. Welcome suggestions for practical procedures from the staff. They can be of great help. For example, the CCU staff will be aware that a CCU patient with a 24-hour

Holter ECG will give better registrations if the electrodes are not put on the back of the patient, because the patient will move during the night. If the staff has to prepare infusions, then provide them with bottles with the right amount of dissolving fluid. Thus, the staff only has to add the medication to the bottles.

4. Make adequate "work schedules"—work plans with clear, step-by-step instructions. Provide boxes that should be ticked off instead of open questions.

NO: Are there any abnormalities at the infusion area?

YES: Tick off any of the following abnormalities observed at the infusion area:

☐ Red

☐ Hot

☐ Inflamed

☐ Swollen

☐ Other, specify: _____

A more accurate and adequate answer is thus more likely to be given. Examples of work schedules can be found in the Appendix to this chapter.

5. Stress that the staff follow the scheduled times as closely as possible. If they cannot meet the schedule, they should fill in the actual time each procedure was preformed, even if its preformance is overdue. The next time they will automatically try to perform the procedure closer to the scheduled time. Moreover, giving compliments really does help.

6. Tell the staff that accuracy is the most important factor in determining the performance of the trial. Ask them to write down everything they do, including extra medication or extra procedures given/performed and so on. If possible, they should also give an accurate indication of the time when the procedure was done.

7. Make absolutely sure that there is always somebody available to address their problems. Provide a list with telephone numbers of persons who can be contacted in case of problems. Then you can avoid excuses like "I didn't know exactly what to do".

8. Arrange a meeting with those staff members who have worked with the first patients and evaluate the process. Thank them for their effort and encourage them to continue the good work (or suggest improvements). Stress again that accuracy is vital. Listen attentively to their problems and do your utmost to solve their problems.

9. During the course of the trial, keep the staff informed about the progress of the trial, the targets and deadlines, and the interim analyses, if available. They will appreciate this and will increase their involvement and motivation.

10. Good preparation is the most vital part of the trial. There should not be any questions with respect to instructions. The instructions should be evident, clearcut, and "fool proof". Even a person who has not been instructed in the procedure should be able to follow the instructions. This will require a lot of effort on the part of the co-ordinator, but it pays off at the end of the trial.

PREPARING FOR THE MONITOR VISIT

The study co-ordinator in the clinic is the liaison with the study monitor. The study monitor will visit the clinic regularly to perform the following activities:

- meet with the investigator to discuss the status and progress of the study, any problems, and supplies;

- review the investigator's file. All documents must be in the binder and up-to-date. If documents are missing, extra copies should be secured;

- sign and date the monitoring visit log;

- collect the CRFs and any other items for review that should be completed prior to the meeting with the investigator;

- review subject activity status since last visit:
  - number of subjects screened for entry;
  - number of subjects entered;
  - number of subjects completed;
  - number of subjects who withdrew and the reason why;
  - number of CRFs completed; and
  - how the recruitment rate is progressing according to plan.
- ensure that any reasons for withdrawal or treatment termination were documented and that adequate follow-up was provided:
- together with the investigator review the CRFs with their respective source documents (i.e., across the table, monitoring) or by direct access. For pivotal clinical studies, check all CRFs at the 100 percent level;
- all corrections and alterations are to be made by the responsible investigator or co-investigator. If agreed, the monitor can correct and change data that are obvious, such as dates and patient record numbers.
- accurately check the investigator's/pharmacist's
  - device dispensing, receipt, and tracking letter;
  - device accountability forms (per patient), including the CRF, if applicable for the study;
  - device return or destruction procedures; and
  - storage conditions; and
- for studies involving biological samples, check sample labelling (i.e., cross-check with CRF) and arrange for transport.

Once the above activities have been performed, the monitor normally will meet with the investigator again with the objective to

- report the monitoring results;

- discuss discrepancies, other issues, and ask him/her to correct certain entries in the CRFs;

- analyse study progress;

- discuss any concerns the investigator may have;

- discuss adverse experiences, especially serious and unexpected events;

- discuss issues from previous visits or discrepancies that became obvious after data editing; and

- assess the continued acceptability of the study site facilities and staff.

The investigator and/or the designated study co-ordinator could increase the efficacy and efficiency of the monitor's visit by preparing for this visit. At least one day before the actual monitor visit, the study co-ordinator could perform the following activities.

- assess the status and progress of subject recruitment;

- collect all documentation and correspondence, and update the investigator study file;

- check and complete any forms with respect to device stock and delivery to the subjects. The device that will be tested in the trial must be inventoried. Although these tasks are often delegated to the hospital pharmacist, it is essential that the investigator and his/her staff (most often delegated to the study co-ordinator) document the stock and delivery of the devices. The study co-ordinator could assist the study monitor with these tasks and discuss a system for the tracking of the devices. In order to not delay the progress of the trial, the study co-ordinator should know the critical stock of devices and should order new devices from the study monitor;

- contact other disciplines within the clinic and discuss any misunderstandings, problems, and questions that should be communicated to the study monitor;

- assess the completion of the CRFs (primary review);

- complete the secondary review; perform quality control on the completion of the CRF; and

- inventory which other study supplies should be requested from the monitor.

PROBLEM SOLVING IN CLOSE CO-OPERATION WITH THE STUDY MONITOR

During the execution of the study, problems may arise that were not anticipated. It is essential that the study co-ordinator functions as the in-house manager of the study. If any problem occurs, the study co-ordinator should be the first person who starts to analyse the problem. The following stepwise approach is recommended.

1. Identify the problem.

2. Discuss with the persons involved the possible causes.

3. Organise a brainstorming session (perhaps a discussion with only one other person is sufficiently adequate) regarding the possible solutions to the problem.

4. Choose a solution.

5. Execute the solution.

6. Evaluate periodically whether the solution was the true solution for the problem.

Managing the project means staying alert for any unanticipated problems during the execution of the study. An optimal flow of information and honesty are conditions for the successful completion of a clinical study. Good quality clinical research requires that open communication and thorough organisation of the clinical investigation take place. The study co-ordinator and the study monitor play a crucial role in achieving this goal.

APPENDIX. SAMPLE WORK INSTRUCTIONS FOR THE STAFF

1. Acute Atrial Fibrillation

Instruction	Scheduled Date and Time	Actual Date and Time	Values
ECG 2x: atrial fibrillation?		A. |_|_| 1981 ____ hours	
Start of episode of atrial fibrillation:			
Check inclusion and exclusion criteria			
• Does patient use forbidden medication that can be washed out? If yes: Calculate start of the infusion = last intake + 5x half life =	B. |_|_| 1981 ____ hours		
• Does the patient use digoxin? If yes: Calculate start of the infusion = last intake + minimally 6 hours =	C. |_|_| 1981 ____ hours		

N.B. If time B or C is later than 48 hours after time A, then the patient is not eligible for the trial!

Appendix continued on next page.

Appendix continued from previous page.

Instruction	Scheduled Date and Time	Actual Date and Time	Values
Physical examination			
Blood pressure =		____ hours	RR __/__
Inform cardiologist			
Inform CCU: location/personnel			
Ask patient for informed consent			
Apply second needle (preferably green) for infusion and needle for blood sampling			
Take blood + urine			
Determine digoxin level if patient uses digoxin		D. |__|__| 1981 ____ hours	
Transfer patient to CCU			

Continue with worksheet 2A or 2B, depending on time D (D = time of blood and urine sampling). If D is *less than 24 hours ago*, then continue with sheet 2A and take the blood sample at time point −10. Cross off sheet 2B, because it will not be used. If D is *more than 24 hours ago*, then continue with sheet 2B and take the blood sample at time point −30. Cross off sheet 2A, because it will not be used.

2A. Acute Atrial Fibrillation CCU

Use this sheet only if the start of the infusion is less than 24 hours after time D.

	Instruction	Scheduled Date and Time	Actual Date and Time	Values
t = −10	• Switch monitor on		|_|_| 1981	
	• Switch Holter on		___ hours	
	• Prepare infusion			
	• Take blood sample		___ hours	
	• Determine digoxin level, if applicable			
t = 0	Do not start if K⁺ < 3.8 or > 5.5			
	• ECG 2x		___ hours	
	• Blood pressure		___ hours	RR _/_
	• Inspect infusion area			☐ no abnormalities

Appendix continued on next page.

Appendix continued from previous page.

Instruction	Scheduled Date and Time	Actual Date and Time	Values
			□ red
			□ pain
			□ warm
			□ itching
			□ swollen
			□ other, specify:
			_____
• Press "Event" button			
• Start infusion		____ hours	
t = 30	• Blood sample		____ hours	
	• ECG 2x		____ hours	
	• Blood pressure		____ hours	RR __/__
	• Inspect infusion area			□ no abnormalities

Appendix continued on next page.

Appendix continued from previous page.

Instruction	Scheduled Date and Time	Actual Date and Time	Values
			☐ red
			☐ pain
			☐ warm
			☐ itching
			☐ swollen
			☐ other, specify:

In case infusion is stopped early (within 60 minutes), take extra blood sample.

2B. Acute Atrial Fibrillation CCU

Use this sheet only if the start of the infusion is more than 24 hours after time D.

	Instruction	Scheduled Date and Time	Actual Date and Time	Values
t = −30	• Take blood sample • Determine digoxin level, if applicable		____ hours	
t = −10	• Switch monitor on • Switch Holter on • Prepare infusion		|_|_| 1981 ____ hours	
t = 0	Do not start if K⁺ < 3.8 or > 5.5 • ECG 2× • Blood pressure		____ hours ____ hours	RR __/__

Appendix continued on next page.

Appendix continued from previous page.

Instruction	Scheduled Date and Time	Actual Date and Time	Values
• Inspect infusion area			☐ no abnormalities ☐ red ☐ pain ☐ warm ☐ itching ☐ swollen ☐ other, specify: _____
• Press "Event" button • Start infusion		____ hours	
t = 30 • Blood sample • ECG 2×		____ hours ____ hours	

Appendix continued on next page.

Appendix continued from previous page.

Instruction	Scheduled Date and Time	Actual Date and Time	Values
• Blood pressure • Inspect infusion area		_____ hours	RR ___ / ___ ☐ no abnormalities ☐ red ☐ pain ☐ warm ☐ itching ☐ swollen ☐ other, specify: _____

In case infusion is stopped early (within 60 minutes), take extra blood sample. At time points 60, 180, 360, and after t = 24 hours, measurements will be performed by the staff of the CCU. Clear step-by-step instructions are given on the worksheets.

21

Data Management

Henriët E. Nienhuis

Parexel/MIRAI
Amsterdam, The Netherlands

Somewhere within the process of data collection, data storage, and data analysis is an element called data management. An exact definition of data management cannot be given, but data management can be described as the process that fills the gap between data collection and data analysis. The purpose of data management is to ensure that the data as presented in the report for each clinical study reflect clinical observations. In other words, data management makes patient evaluations suitable for statistical analysis.

In the ideal case of perfect patients who fulfil all protocol criteria without any deviations, data management might be a waste of time. However, even in this hypothetical case, data management can be very useful, as the tools of a data manager are very helpful for performing a clinical study efficiently.

TIP	In addition to clinical personnel and statisticians, it is very important to have data managers involved from the start of the project.

Because data managers are experts with regard to computerised data, they will know exactly which kind of data will help to answer the objective(s) as formulated in the protocol. In other words, for the operationalisation of the objective(s) and the translation of the data into case report forms (CRFs), data managers are of great value. In general, the process of data management of a medical device trial does not deviate significantly from the data management process in a pharmaceutical clinical trial. In this chapter, the data management part of the clinical trial process will be explained, as well as what the impact of data management is and how the process of data management works.

DESCRIPTION OF THE DATA MANAGEMENT PROCESS

The handling of data collected in a clinical trial is a job that should be done by specially trained data managers. They need to organise, structure, and check the data in order to ensure it is in good shape for statistical analysis.

The data management process can be divided into three parts, which will be described in the following sections. The first part of a clinical trial in which the data manager should be involved is the design of the CRF. As the data manager is very familiar with all kinds of data, this person will know exactly what the different options are with regard to the design of the CRFs.

Data Entry

Data entry can be regarded as the first part of the data management process. At first glance, data entry is a process that looks as if it is a very easy job. In fact, it often is time-consuming and difficult; it requires entering data into the computer without errors. The first task is to learn to recognise when data have been entered correctly. A guarantee of 100 percent certainty can never be given. However, several methods and systems are available to achieve near-certainty. Because systems and methods are closely related to each other, as will be explained, one needs to know from the start what the objective is. To do this, one first has to establish the number of errors that will be acceptable. In reaching the objective, one can select from the following options.

Single Data Entry

In single data entry, all data are entered once. This is, of course, the quickest and cheapest way of entering the data into the computer. However, there will be no check for typing errors. Of course, although typing errors do not necessarily result in major consequences, they often do. One can imagine that when a typing error is made in, for example, a patient number, the data can be shifted to another patient erroneously, which can have a major impact on the outcome of the study. One can therefore state that it is very important to have qualified data entry personnel who have been trained adequately. Project-specific training should be documented when using this method. This method, however, is not preferred.

Single Data Entry with Verification

With this method, the data are checked after they have been entered. This checking process can be done manually, but there are also software packages available for automatic, computerised checking, such as ClinTrial™. An example of a manual check can be the following: All data from CRFs that are entered will be printed and independently visually checked by someone other than the data entry person involved.

ClinTrial™ is a computer data package which can be used for data entry and verification of the entered data. ClinTrial™ allows data to be entered a second time. There are two alternatives within the programme. The first requires entering the data for the second time; the computer will produce a beep when the data entered differs from the first entry. The second data entry is done by an individual different to the person who did the initial data entry. In case of data difference, the second data entry person decides which of the two entries is correct. The second alternative is that the items for which the second entry deviated from the first entry are written to a so-called "discrepancy list". This list then has to be checked with the data as recorded in the CRF. In the optimal case, this list is checked by someone other than the primary and secondary data entry persons. Subsequently, the appropriate corrections have to be made. The advantage of this system is that only one database has to be corrected, as only the first data entry set will be saved; for the second data entry set, only the discrepancies will be noted. A disadvantage is, at least in the first example, that the second data entry person has to decide which of the two data sets is acceptable. This makes the second data entry set of greater importance than the first. Furthermore, since only the first data set is stored, it is important that it be of a high quality; this will save corrections afterward. Although this method is preferred to single data entry, one needs to be careful to select primary and secondary data entry persons who are well qualified.

Double Data Entry

As can be concluded from the name, double data entry is the data entry process whereby the data are entered twice, into two physically separate databases. Afterward, the two databases are compared by using, for example, the SAS procedure Compare. Output of this procedure generates a listing that indicates the differences between the two databases. When this listing is compared with the CRF, it can be determined which entries in each of the two databases are correct and which need to be corrected. Thus, both databases are corrected when necessary, and the comparison procedure is repeated until the two databases are identical and no output is produced by the comparison procedure. Besides being the most reliable method, this method is also the most time-consuming and expensive method. However, this method is preferred over the previous methods described.

Data Validation

After the data have been entered, they need to be validated. The process of data validation is meant to ensure that the data that have been entered are correct. This can be done by using trained monitors, but also by using computerised checks specifically programmed for the particular study. These checks are often referred to as *validation checks* or *error checks.*

The checks are merely designed to check for inconsistencies that have occurred in the CRFs, because they can cause problems when the data have to be statistically anlysed. The data validation process can be roughly divided into three parts: the medical part, often called *medical review;* the range checks; and the consistency checks.

Computerised checks are more advantageous than manual checks. However, although computers will check data in a consistent manner and will not overlook things, it is very unlikely that one can design a clinical study in such a way that all checks can be performed by computers. The process of medical review, for example, will be a nearly impossible job for a computer, while consistency and range checks can usually be computerised. Validation checks should always be study-specific, though some general checks are useful in every study. Examples of general checks include checks on the date, checks on the patient number, and checks on patient inclusion and exclusion criteria. A further example would be a check of the correctness of the patient number, which should be the same on every page of the CRF. Another example is the visit date; if visits are scheduled every week, the interval between the visits should be close to 7 days. Of course, the date of any visit also has to be after the date of the previous visit.

When the data have been monitored accurately, the validation checks will not produce any output and no queries will be generated. As the query process is not only very time-consuming, but also error-sensitive, the occurrence of queries should be avoided. This may look like a very difficult job, but in practice, it has been shown that involving the data manager in CRF design is very helpful.

Documentation of the data management process has to be done properly. This is accomplished by producing a data handling manual. In this document, the whole process of data entry, data verification, validation, and clarification is described. The purpose of a data handling manual is to document the process of data handling in a detailed way, making it understandable for everyone involved in the process. Consequently, when the data handling manual is written correctly and adequately, the data processing and management part of a clinical study can be taken over by anyone familiar with the techniques used. The contents of a data handling manual should include the handling of the data from its arrival at the data centre up to and including the delivery of a clean database to the statistical department. Not only data administration should be described, but also data entry. In addition, the manual should have a section of data entry rules that are updated during the study. This ensures that the handling of special cases is done in a standardised manner. Furthermore, the systems used for data entry as well as the persons responsible should be mentioned. Any special data management tool that is used (such as DataFax™) needs to be documented in terms of its deviations and in its comparison to the "normal" data management process.

Not only the systems to be used but also a description of the database in terms of field widths, variable names, and formats (i.e., a codebook) should be given as well as a detailed description of the validation checks to be performed by the computer. This enables the monitor to verify which items will be checked by the computer; one can also obtain a good overview of the quality of the data to be expected in relation to the checks performed.

WHEN TO START DATA MANAGEMENT

The importance of having data managers involved early on in the process of CRF design cannot be stressed enough. Once the study is running and patients are included, data management should start as soon as the first patient data are entered into the computer.

It is worthwhile to put every effort into the design and programming of the validation checks before the first data are collected.

TIP   The designed validation checks should be discussed with the monitors involved in the study. This should be done during a special training session for all monitors. This will lead to a standardised method of data monitoring, a better understanding between monitors and data managers, and higher quality of the validation checks. When monitors are aware of the checks that are to be performed, they can pay extra attention to these items during their monitoring activities.

It is worth trying to have the first validation checks running directly after the first data have been entered. This way the results can be verified directly and the validation checks can be adapted. In addition, it is also possible to correct investigators' errors at an early stage, which prevents additional errors of the same type from occurring and saves time and money.

Moreover, an ongoing process of data validation and *cleaning* will result in only a short period of data entry after the last patient has completed the study. Similarly, due to data cleaning, statistical analysis can start shortly after all patients have been enrolled in the study, instead of weeks later.

Smart programmers can also design error checks that will result in automatically generated queries, which, when neatly printed, can be sent directly to the investigators. This process will involve a lot of time and effort at the beginning of the study, but will be of great value afterward. It is also important to decide on the standardisation of CRFs and the database supporting it. When performing more than one study, the general checks described previously and checks on standard items (e.g., inclusion and exclusion criteria, demographic data, a standardised CRF, and accompanying database) can then be used for all following studies without major effort.

HOW TO MANAGE DATA

The data management process should start as soon as data collection has started. Depending on the way the data are collected, the data flow can be organised as described below. As soon as the data arrive at the data centre, the data management department should start performance checks of the data. All incoming data have to be administrated and locked in a safe as prescribed by the rules of Good Clinical Practice (GCP) in order to ensure the complete filing and tracking of all pages in a CRF. Thereafter, the checking process begins. Initially, a number of general checks will be performed; in most cases, this is easiest during the administration of the data. The incoming data must be checked for completeness: Are the data for each patient complete? Are there parts of the CRF that have not been completed? Should this be the case, it should be verified and confirmed that data is actually missing. The visual checking, as performed by the monitor, is an extra check to ensure that there are no data available that have not been recorded in the CRF.

After the first check and administration, the data should be transferred to the data entry department, where they are entered and verified. Subsequently, the validation process starts. After all checks have been performed, queries are generated. An example of a query form is given in Figure 21.1. The queries have to be answered by the investigator, which may cause changes to the data entered. Of course, the querying process will undergo the same routine as previously described, so that all queries will be resolved or confirmed. A graphical presentation with regard to the data handling process as described earlier is presented in Figure 21.2.

At the end of the trial, there must be absolute certainty that all of the data collected in the clinic have been entered. Once all of the data have been collected and entered and all queries are resolved, the data can be regarded as clean. Technically, it will then be the responsibility of the data manager to close (i.e., 'lock') the database. In this way, modifications of the data are no longer possible and the data are ready for statistical analysis. The milestone of obtaining a clean database should be documented properly.

Often, the question is raised about what quantity of queries is acceptable. Basically, a straightforward answer to this question cannot be given, as it depends on a number of factors. First of all, it is important to know to what extent the data are checked. In general, more checks will result in more queries,

Figure 21.1. Query Form for Data Entry

Protocol No.		QUERY FORM	Patient Initials	Query raised by:
Study Device:			Site No.	Date:
Investigator name			Name Monitor	

Query No.	CRF No.	Item No.	Query	Resolution	Code	On DB 1st	On DB 2nd

Signature Monitor: .... Date:

Signature Investigator: .... Date:

Resolution Codes:
1. Signed off by monitor
2. Reference to Source Data
3. Reference to other documentation (e.g. fax, letter)
4. Signed off by investigator

Query form No.: --------------------

Figure 21.2. Data Handling Process

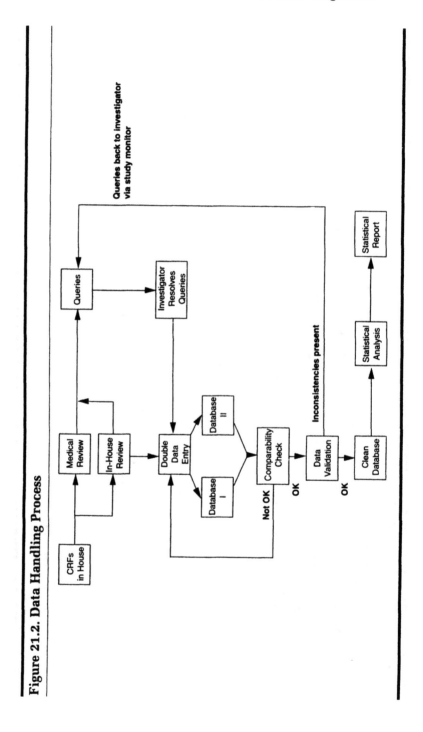

improving the accuracy of the data. However, a large number of queries is not preferred, as this is a time-consuming and error-sensitive process. Thus, one needs to find a balance between the accuracy desired and the efforts to be exerted.

| TIP | It is possible to identify certain variables as key variables, for which a high level of accuracy is desired, whereas other variables will be of less importance. Consequently, the number of checks on the key variables can be larger than on the variables of less importance. |

Another factor of influence is the justification for sending a query to the investigator. In this respect, we can distinguish between explanatory queries, self-evident queries, and confirmative queries. Explanatory queries are queries that represent discrepancies in the data collected; thus, they always need to be answered by the investigator. In the case of self-evident queries, (which occur when it is evident that data collected in the CRF is inconsistent with what the answer to the CRF question should be) one can decide to transform it to a confirmative query. In general, this type of query is less error-sensitive, from a data management point of view. Confirmative queries are queries in which the investigator is asked for a confirmation. For example, one could think of spelling mistakes that need to be corrected. This implies that this type of query can be solved by people other than the investigator, for example, the monitor involved. This means, however, that when self-evident queries have been transformed to confirmative queries, self-evident queries can also be solved by the monitor. This illustrates a grey area and one should be very careful about letting monitors resolve queries. In all cases, it should be documented who is allowed to write on the CRF.

| TIP | Should it be decided to allow monitors to write on the CRF in order to prevent confirmative queries, the monitor should use a different colour pen, and initial and date the data added, changed, or clarified. This gives traceability and documents what has been done by whom. |

Last, but not least, the working procedure with regard to the data cleaning process is of importance when considering the acceptable number of queries. When using no carbon required (NCR) paper, two methods are widely used.

1. The so-called working copies are directly data entered, without monitoring. This has implications for the validation process. The checks to be performed by the computer will have to be very accurate and extensive, as the output from these checks will be of great importance in the cleaning process. Of course, the number of queries will then be extensive.

2. Use the working copies for monitoring purposes. The data will then be entered after the monitor has not discovered any discrepancies. This means that the validation process is an extra check on the data to ensure that the monitor did not overlook things. Consequently, the number of queries resulting should be minimal.

DATA MANAGEMENT IN RELATION TO OTHER DISCIPLINES

The complexity of the data management process depends on a number of factors. Of course, the design and the objective of the study have impact, but logistic factors can also be distinguished. In planning a trial, it is important to take into account the impact of such factors on the data management process, as this not only improves the quality of the trial, but also saves much time.

Besides a number of study-specific decisions, a couple of general issues should be noted. First of all, it is important to know where the study will be carried out. Multi-centre studies are more complex than single-centre studies; the consequences for data management are numerous. In the more complex situation of a multi-centre trial, it is very important to have a standardised method for data handling, an item that is often overlooked. In the ideal case, one person should be responsible for the data handling at all trial centres. However, due to practical reasons, this is often not possible. Preferably, each centre should have its own data handler or monitor; it is highly recommended to train all of the data handlers together, not only before the enrolment of the first subjects, but also during the course of the trial.

Secondly, in case the study will be carried out in more than one centre, it is important to know whether this also means that it will be an international study. When this is the case, it has implications for the procedures of data management, as each country will have its own specific regulations. Differences among countries will cause problems during data entry, as the database to be used will have only one format. The data entry personnel cannot be expected to understand all of the languages used.

TIP Decisions on the units to be used should be made, as this will have implications for the format of the database. In case patient diaries are to be used, these should, of course, be in the local language. An important note should be made: In the case of diary cards in different languages, it is very important that they be checked very carefully to ensure that the different diary cards are exactly the same in each language. The layout should be uniform, and the sequence of the questions the same.

Another item of importance with regard to the logistic factors is the way the data are collected. This can be done on a per-patient basis; a visit-by-visit approach is also possible. The advantage of collecting "whole patient" data is that the data will arrive at the data centre in one piece. It can, therefore, be administrated and processed quickly. All queries raised can be handed to the monitor and investigator at one time, which makes unnecessary a continuous administration of outstanding and unresolved queries. This makes the process of data management transparent and straightforward. However, the disadvantage is that the whole process of data management is postponed until the end of the trial. This means that the time between the last subject completing the study and the closure of the database will be relatively long. Furthermore, the possibilities for steering and training the investigator during the study are limited.

On the other hand, visit-by-visit collection of the data has the advantage that it is possible to respond to the investigator quickly and to clean the data on an ongoing basis. This minimises the length of the cleaning period after the last patient has completed the study. However, this approach requires a very accurate and time-consuming administration of the data that are being processed, the queries outstanding, and so on. Clearly, it is preferable to simplify the study with regard to these factors as much as possible.

In relation to the monitoring of a clinical study, it is very important to have regular discussions between the data manager and the monitor. In this way, the monitor will have a better insight into why queries have been raised and will be able to prevent queries in the future. On the other hand, the data manager will be more informed on the clinical aspects of the study, and it might be possible to adapt some data checks or pay some extra attention to items that have proven to cause problems.

TIP	In the case of a multi-centre study, such discussions can be organised in training sessions. Training sessions are also useful in obtaining a standardised method of data handling.

Furthermore, the statistician and data managers should work together. This will prevent problems during data analysis. All special cases should be discussed, and the data manager and statistician should decide how to handle these data. In this respect, the statistician will already have some familiarity of the data before the analysis starts.

CRF DESIGN FROM THE DATA MANAGEMENT POINT OF VIEW

A couple of guidelines for the design of a CRF are useful. First of all, one has to distinguish between the type of data, numeric or character. Numeric data consist of numbers, whereas character data contain text. A data manager will always prefer numerical data, not because he/she is too lazy to read, but simply because computers can more easily work with numbers than with characters. This also indicates that numerical data are more suitable for performing computerised checks. One can also easily imagine that typing errors are more often made in text (i.e., spelling mistakes) than in numbers. Moreover, the fact that entering numbers can be done much more quickly than entering text is also a valid argument for preferring numerical data over character data.

Second, the investigator should not be invited or allowed to write in the margin of the CRF. To discourage the use of unsolicited text, it is important that the questions on the CRF actually fill the page, (i.e., there should be no room for the investigator to write comments). However, a comment section at the end of each visit can be very useful, as the comments are written together and can easily be retrieved when necessary.

Third, it is very important to make the CRF self-explanatory. The investigator has to be able to fill in the CRF without any instruction. Of course, it will be necessary to provide instructions for some items. In that case, the following tip is useful.

TIP | The backside of each CRF page can be used as an instruction page for the completion of the facing page. A few examples of different situations noted on the back of the facing page are usually very instructive.

The CRF should be built up in a logical manner. Thus, anamnestic data should be grouped together. It is also very efficient, not only from the investigator's point of view, but also in terms of data entry, that pages concerning different visits contain the same type of data and have the same layout. Depending on the way the data will be collected, visit-by-visit or as patients complete the study, it is useful to have identifying information (e.g., the patient number) on every consecutive page of the CRF.

TIP | To prevent mistakes in filling out the patient number and the subsequent, disastrous entry of the error during data processing, one can decide to have the number printed on every page of the CRF. Furthermore, it is absolutely necessary to have only one unique, identifying number per patient. Thus, there should be only one patient No.1.

Another important point in data collection is the objective of the study. Almost everyone seems to have a tendency to collect as much data as possible and with as much detail as possible. However, one should always have in mind the objective of the study, which data are absolutely necessary to answer the objective, and how much detail these data require. For example, in a study designed to investigate the accuracy of blood pressure measurements, it might be useful to know whether or not the patient has high blood pressure, because it will be interesting to know whether the device under investigation is more or less sensitive when measuring certain blood pressure ranges. A question in the CRF to record this information does not have to contain information regarding the duration of the current (high) blood pressure, the reason, nor all of the medication used in the past with regard to this condition, but it could simply be a yes/no question: "Does the patient have high blood pressure (defined as . . .)?" In this way, the answer to this question provides all the information needed, while no data are collected that cannot be analysed and that do not contain useful information.

TIP	Questions must be relevant to the protocol.

In addition, it is also very important to pay attention to the structure of questions in the CRF. Questions should be 'closed' as much as possible, and there should not be any negatively phrased questions, because they are often misread. There should only be one question asked at a time, so that the answer is always clear.

TIP	When the investigators are instructed to send the working copies of the CRF to the monitor directly after each visit, the data can be monitored without the monitor having to be on-site. This might save (travelling) time, and also provides the opportunity to respond to the data immediately, if necessary.

In addition to the CRF design, it has to be decided what material is to be used for the CRF. NCR paper is widely used.

This type of paper has both advantages and disadvantages. As described previously, NCR paper allows the use of one copy of the CRF as a working copy.

Another advantage of the use of NCR paper is that the investigator will have his/her own copy, and photocopying of the CRF after the completion of the trial in order to provide the investigator with a copy of the patient data is no longer necessary. However, it is absolutely necessary to use high quality NCR paper, to make sure that the last copy of each page is still readable.

On the other hand, once the original copy of the CRF page has been removed and corrections are made, these corrections must also be made on the top copy of the investigator's copy. In case ordinary paper is used for the CRFs, this step will not be necessary, as these pages are to be photocopied once all data are final and clean. In conclusion, the use of NCR paper is preferred when the working copy of the CRF is used intensively.

22

Filing and Archiving

Herman Pieterse

Profess® Medical Consultancy BV
Heerhugowaard, The Netherlands

In order to conduct a study in compliance with U.S. and European Good Clinical Practice (GCP) guidelines, the study must be verifiable. All documents needed for an audit or for an inspection must be retained intact and complete, and must be easy to find, even many years after the study has been completed.

Companies sponsoring clinical studies should write a standard operating procedure (SOP) concerning the filing and archiving of documents. In this procedure, instructions should be given concerning

- the active study file;

- modalities for archiving at the sponsor company (i.e., the Trial Master File [TMF]);

- the content of the sponsor's archives; and

- the investigator's archives.

Archiving clinical trial documents is an essential part of the clinical trial process. The protocol, documentation, approvals, and all other documents related to the trial, including certificates demonstrating that satisfactory audit and inspection procedures have been carried out, must be retained by the sponsor in the TMF. The clinical trial co-ordinator is the primary person responsible for maintaining proper filing and archiving of all documents and correspondence.

The study monitor ensures that the investigator arranges for the retention of patient identification codes. Subject files and other source data must be kept for the maximum period of time permitted by the hospital, institution, or private practice, according to local law. The sponsor needs to retain all other documentation pertaining to the trial for the lifetime of the product. Archived data may be held on microfiche or electronic record, provided that a backup exists and that hard copy can be obtained from it, if required. All relevant documents and correspondence are to be filed in an investigator study file or clinical trial binder.

Prior to the initiation visit, the investigator study file should be prepared for each site. This file is prepared by the sponsor. The monitor needs to ensure that the investigator maintains a complete and accurate record of the clinical study conducted and provides the investigator with the means to do so. The investigator must have a complete and up-to-date investigator study file. A copy of the documents in this file are also to be retained at the sponsor's site.

THE ACTIVE STUDY FILE

During the preparation and execution of a trial, many documents are used to collect information and to support conformity with quality procedures. This information need to be accessible to users. These users may be the project directors, monitors, clinical research associates (CRAs for both data and monitoring), the shipping department (for device monitoring), and the statistical department (for data processing). These documents should also be available to and accessible for an audit and an inspection.

During the course of the study, these documents should be indexed and classified according to a filing plan. Access to the documents should be limited to users who are clearly designated by another procedure. The locking of filing cabinets in facilities close to users' offices should be done.

WHAT ESSENTIAL DOCUMENTS SHOULD BE FILED?

The following documents should be collected prior to, during, and after closure of the study:

- Investigator's Brochure or equivalent documents;

- final protocol (signed), (signed) amendments, if applicable, and Operations Manuals (investigator handbook, laboratory manual, and/or monitor manual);

- blank case report form (CRF) (sometimes found in the protocol) and a copy of the completed CRFs;

- curricula vitae of the investigator, co-investigator(s), and research assistant(s);

- a list of names with signatures and initials of all study staff who will fill in, sign, or make corrections to CRFs;

- documentation from the Institutional Review Board (IRB) or Ethics Committee (including approval of protocol, informed consent procedure, and correspondence);

- contract and/or financial documents. (If the contract is signed directly by sponsor and investigator, it is the responsibility of the sponsor to have the document in the study file.);

- signed letter of confidentiality (if applicable);

- signed letter of agreement (if applicable);

- blank consent form and patient information sheet (signed consent forms from the patients should remain with the source or be archived in the investigator study file [ISF]);

- laboratory normal values, methods (if available), and lab certificate;

- pre-study visit and study initiation visit reports;

- device supply, return, and accountability forms;

- all correspondence (e.g., letters, telephone call notes, etc.) pertaining to the study (sites have to only keep the correspondence to and from them on file);

- a copy of the monitoring visit log;

- blank adverse event forms and a copy of the completed forms; and

- patients' randomisation envelopes (if applicable).

TIP | The financial documents, official contracts, financial agreements, and the signed protocol plus amendments could be filed with the director's secretariat and not in the study file. Then a reference should be included in the study file stating where an auditor can find these confidential documents.

HOW TO SET UP A TMF

The study project manager (i.e., CRA or study monitor) sets up the file system. This could be done, for example, by using a binder or subject cards. Because each clinical trial differs in complexity, the project manager is not obligated to follow the same order or subdivision for each TMF. However, the project manager is responsible for filing, maintaining, and updating the TMF, and should keep all of the following:

- product-related documents (e.g., Investigator's Brochure);

- project-related documents (e.g., study protocol, CRF, informed consent, manual of operations, study reports,

Ethics Committee/IRB approval letters and correspondence, etc.);

- contracts and agreements (e.g., letter of agreement, contract, curricula vitae of the investigators and co-investigators, contract with the sponsor, etc.);

- correspondence and reports to and from the investigators (per investigational site);

- correspondence to other parties;

- minutes of meeting, memoranda; and

- patient-related documents (e.g., the CRFs).

The project manager should ensure that these respective documents are properly stored and kept up-to-date.

RETENTION OF RECORDS AT THE INVESTIGATIONAL SITE

The investigator is responsible for ensuring that all subject records as acquired during the trial are archived and available for inspection by authorised sponsor personnel or regulatory authorities. The subject files should be retained for the maximum period of time permitted by the hospital, institution, or private practice. Should the investigator or the hospital not guarantee to archive the documentation for the required period, the project manager should discuss alternate arrangements with the sponsor. The investigator's files, regardless, are to always remain under the control of the investigator. Should the investigator be unable to maintain custody of the trial documents, the project manager needs to be informed in writing concerning the location of the records and the identity of the person(s) responsible for their retention.

TIP	The project manager could arrange the study file to be archived by an external archiving company. These companies render services for securely maintaining archives. Access to the files is permitted to authorised persons only.

HOW TO SET UP A STUDY FILE AT THE INVESTIGATIONAL SITE

The study monitor is the primary person responsible for setting up and updating the study file. The study needs to ensure the following are undertaken.

- During the study initiation visit, the study monitor discusses the documents with the investigator that are to be filed at the investigator site and offers assistance in establishing a filing system.

- During the study initiation visit, the study file binder will be presented to and discussed with the investigator, and the filing system will be explained, i.e., who has to do what by when.

TIP	The monitor should explain to the investigator that he/she and the study monitor may add documents to the study file. The monitor reviews the file to ensure that it contains all of the essential elements for inspection by authorities.

- Review with the investigator any regulations for retaining study records:

  - Investigator study file, for instance, including, patient identification code list and signed informed consent forms, which per clinical practices laws could have to be retained for a period of a minimum 15 years after the completion or discontinuation of the trial.

  - Patient files are to be retained for the maximum period permitted by the hospital or practice.

  - Sponsors are to retain all trial documentation, for the lifetime of the device.

- Explain to the investigator that copies of the CRFs, the source documents, and signed informed consent are to be retained.

- Explain to the investigator that the sponsor is to be kept informed of any change of address or other relocation of study files.

- Specify in the study initiation visit report where the investigator will store office charts, clinic records, or other raw data for study subjects.

WHO HAS ACCESS TO THE STUDY FILES?

The sponsor should define the authorisation of access to the archives in writing, listing who may consult, correct, add, copy, move, borrow, or destroy documents. Borrowing or destroying documents should be subject to special precautions. In the ideal situation, in order to avoid the removal of any original documents, photocopies need to be made on-site. It is preferred that each time the archives are consulted, a dated and signed form should be completed. In practice, this is rarely possible. It is the responsibility, however, of the study project manager to keep the archives accurate and complete.

TIP When documents must be sent by post from the investigational site to the sponsor, make sure that this occurs via a courier service, and not via ordinary post. Make a copy of any documents sent. In practice, the monitor transfers documents from one file to the other by hand.

HOW TO CLASSIFY THE ARCHIVES

Archived documents must be easy to consult. The procedure adopted most often is a chronological classification of complete study files (with the contents as proposed earlier in this chapter), including all company products tested. In multi-centre trials, the company must plan a general classification file (a TMF) containing all documents relevant to all centres, and, a file for each individual centre. A register should be established in order to find the location of the shelves; if archives are voluminous, a computerised system may be useful.

HOW LONG SHOULD THE ARCHIVES BE KEPT?

The legally required duration of archiving (accessible to inspection) varies from country to country. In all cases, documents should be retained in accordance with either regulatory requirements as deemed appropriate, or in accordance with wishes of the regulatory authorities. Device clinical trial files must be retained in relation to the life-time of the device. For example, for implantables this may be the lifetime of the implantee, (i.e., permanent). Other elements influencing duration/retention of files are liability laws and the sponsor organisations' internal file retention policies.

It is advisable not to destroy archives, but to store them in a secondary archiving facility. Indeed, other uses may occur: responsibility in the case of sequelae or late-onset complications, review of the validity of the intended purpose, or regulatory approvals, or emergence of long-term complications. Some companies require archiving for 30 years or more, others for 10 years.

NEW TECHNOLOGY TO STORE
AND ARCHIVE DATA

Voluminous documents may be stored on microfiche or on videodiscs that can store page images. Specific software is commercially available that can index, image, and store each document. The acceptance of this type of archiving by regulatory authorities is, of course, one of the first factors that needs to be considered when investigating this option. Adoption of these techniques requires visual-display monitors adapted to the frequency of consultation of the archives. Obtaining paper copies of information in case a thorough examination is needed should still be possible.

Computer files may be stored on floppy discs or magnetic tape. In either case, they must be periodically recopied (in order to verify data restoration). This disadvantage is eliminated with laser discs for computerised data (for single recording and multiple displays), but the equipment needed is expensive.

AFTER STUDY COMPLETION ARCHIVING

Study archives should be placed, where possible, under the responsibility of an "archivist". He/she organises, establishes, and maintains the filing. The archivist also verifies that applicants who wish to consult the archives are indeed authorised to do so. The archivist is also responsible for removal and replacing of files consulted to avoid loss or misfiling. If the archivist is absent, a replacment needs to be designated.

APPENDIX A. INVESTIGATOR'S STUDY FILE CHECKLIST

The sample contents given below of an investigator's study file can be used as a checklist and for guidance at the front of the binder.

Content	Check/Date
Communication	
• Visit log	
• Investigator meetings	
• Correspondence	
Subjects	
• Subject identification log	
• Signed informed consent/patient information sheets	
• Other documents	
Supplies	
• Study devices	
• Random code envelopes (during the study)	
• Study materials	
Laboratory	
• Central laboratory	
• Local laboratory	
Investigator Documents	
• Authorised personnel form	
• Curricula vitae of investigator and other study personnel	

Appendix A continued on next page.

Appendix A continued from previous page.

Content	Check/Date
• Investigator's statement	
• Confidentiality agreement	
• Receipt of Investigator's Brochure/ product information	
• Financial contract	
Ethics Committee/Regulatory Affairs	
• Ethics Committee approval(s)	
• Ethics Committee approval(s) of amendments	
• Ethics Committee members' (voluntary)	
• Correspondence with Ethics Committee	
• Insurance/indemnity certificate(s)	
• Legally required documentation, if required	
• Correspondence with regulatory authorities	
Adverse Events	
• Adverse event reports	
• Correspondence	
Study Documents	
• Address of sponsor	
• Address of principal investigator	
• Address of contract research organisation	
• Blank informed consent/patient information sheet	

Appendix A continued on next page.

Appendix A continued from previous page.

Content	Check/Date
• Investigator's handbook (part of Operations Manual)	
• Protocol (including signature page)	
• Protocol amendment(s) (including the signature page of the amendments)	
• Investigator's Brochure/product information (and updates)	
• Guidelines and laws	
Case Report Forms (separately filed)	
• Completed CRFs including query forms	
Final Study Report	
• Final study report	

APPENDIX B. TMF CHECKLIST

Content	Yes	No	Date/ Initial
List of Centres			
• Participating centres			
• Planned but not participating centres			
• Documentation			
Responsibilities			
• Allocation of tasks, with signatures			
• Curricula vitae of sponsor staff (filed centrally)			
• Allocation of investigators to monitor			
Study Documents			
• Protocol*			
• Protocol amendment(s)			
• Informed consent/patient information sheet*			
• CRF			
• Other study-specific documentation			
• Investigator's Brochure (and updates)/product information			
• Operations Manual(s)/ investigator's handbook*			

*Draft versions and draft translations should be filed in a separate binder.

Appendix B continued on next page.

Appendix B continued from previous page.

Content	Yes	No	Date/ Initial
Ethics Committee			
• Approval(s), if applicable			
• Approval(s) of amendments, if applicable			
• Members' (voluntary)			
• Insurance/indemnity certificate(s)			
Regulatory Affairs			
• Legally required documentation, if applicable			
Supplies			
• Study devices			
• Emergency code envelopes			
• Study materials			
• Randomisation procedure, if applicable			
• Correspondence			
Laboratory (central)			
• Curriculum vitae of director of laboratory			
• Reference ranges			
• Certificate(s)			
• Contracts			
• Correspondence			

Appendix B continued on next page.

Appendix B continued from previous page.

Content	Yes	No	Date/ Initial
Serious Adverse Events			
• Overview			
Communication			
• Communication with all involved parties			
Meetings			
• Investigator meetings			
• Internal meetings			
Status			
• Reports			
Quality Assurance			
• Audit reports/certificates			
• Follow-up of audits/follow-up reports of audits			
• Correspondence			
Data Management**			
• Data handling manual			
• CRF tracking log			
• Data entry			
• Data validation			
• Coding			

**Documents filed by the responsible person/department. The monitor should file the documents for a specific centre in a separate centre-specific file, which is an integral part of the TMF.

Appendix B continued on next page.

Appendix B continued from previous page.

Content	Yes	No	Date/Initial
Statistics**			
• Statistical analysis			
• Randomisation, if appropriate			
Report			
• Report			
• Publications			
Finances**			
• Financial agreements/contracts			
• Internal budgets			
• Invoices/payments			
• Letter of intent			

**Documents filed by the responsible person/department. The monitor should file the documents for a specific centre in a separate centre-specific file, which is an integral part of the TMF.

APPENDIX C. CENTRE-SPECIFIC FILE CHECKLIST

Content	Yes	No	Date/ Initial
Monitoring			
• Communication			
• Contact reports			
• Reports			
• Correspondence			
• Documents necessary for the preparation of the next visit, such as agenda			
Protocol/Amendments			
• Protocol signature page			
• Amendment signature page(s)			
Ethics Committee or IRB/ Regulatory Affairs			
• Ethics Committee or IRB approval(s) (local)			
• Ethics Committee or IRB approval(s) of amendments			
• Ethics Committee or IRB members (voluntary)			
• Correspondence or IRB with Ethics Committee			
Investigator Documents			
• Authorised personnel form			
• Curricula vitae of investigator and other study personnel			

Appendix C continued on next page.

Appendix C continued from previous page.

Content	Yes	No	Date/Initial
• Investigator's statement			
• Confidentiality agreement			
• Receipt of Investigator's Brochure/product information			
Supplies			
• Study devices			
• Random code envelopes, if applicable			
• Study materials			
Laboratory (local)			
• Address			
• Curriculum vitae of director of laboratory			
• Reference ranges			
• Certificate(s)			
Patients			
• On-site patient log			
• Source documents verification forms			
Finances			
• Overview			
• Financial agreement/contract			
Case Report Forms (separately filed)			
• Completed CRFs including query forms			

APPENDIX D. ESSENTIAL GCP DOCUMENTS

Essential documents that are collected in a clinical trial are described below. This table could be used as a guide for setting up and maintaining a study file for a clinical investigation with medical devices.

Before the Trial Commences

Before formal start of the trial, the following documents should be generated and should be on file.

Title of Document	Purpose	Located in Files of	
		Investigator/ Institution	Sponsor
Investigator's Brochure	To document that relevant and current scientific information about the investigational product has been provided to the investigator.	X	X

Appendix D continued on next page.

Appendix D continued from previous page.

Title of Document	Purpose	Located in Files of	
		Investigator/Institution	Sponsor
Signed protocol and amendments, if any, and sample Case Report Form (CRF)	To document investigator and sponsor agreement to the protocol/amendment(s) and CRF.	X	X
Information given to trial Subject			
• Informed consent form (including all applicable translations)	To document the informed consent.	X	X
• Written information for subject regarding the device	To document that subjects will be given appropriate written information (in regard to content and wording) to support their ability to give fully informed consent.	X	
• Procedure for subject recruitment (if used)	To document that recruitment measures are appropriate and not coercive.	X	X

Appendix D continued on next page.

Appendix D continued from previous page.

Title of Document	Purpose	Located in Files of	
		Investigator/ Institution	Sponsor
Financial aspects of the trial	To document the financial agreement between the investigator/institution and the sponsor for the trial.	X	X
Insurance documents	To document that compensation to subject(s) for trial-related injury will be available.	X	X
Signed agreement between respective parties, e.g.:	To document contractual obligations and responsibilities.		
• Investigator/institution and sponsor		X	X
• Investigator/institution and CRO		X	X
• Sponsor and CRO		X	X
		X	X
• Investigator/institution and authority(ies) (where required)		X	(where required)

Appendix D continued on next page.

Appendix D continued from previous page.

Title of Document	Purpose	Located in Files of	
		Investigator/ Institution	Sponsor
Dated, documented approval/favourable opinion of Institutional Review Board (IRB)/Ethics Committee of the following: • Protocol and any amendments • CRF (if applicable) • Informed consent form(s) • Written information to be provided to the subject(s) • Subject procedure recruitment • Subject compensation (if any)	To document that the trial has been subject to IRB/Ethics Committee review and given approval/favourable opinion. To identify the version number and date of the document(s).	X	X

Appendix D continued on next page.

Appendix D continued from previous page.

Title of Document	Purpose	Located in Files of	
		Investigator/ Institution	Sponsor
• Any other documents given approval/ favourable opinion			
Institutional Review Board/Ethics Committee Composition	To document that the IRB/Ethics Committee is constituted in compliance with GCP.	X (where required)	X
Approval/notification of trial from regulatory authorities (if required)	To document appropriate approval/notification by the regulatory authority(ies) has been obtained prior to initiation of the trial, in compliance with the applicable regulatory requirement(s). For commercialised product under clinical trial, copy of certificates of CE-marketing from NB or similar approvals. Include ER's checklist for EU.	X	X

Appendix D continued on next page.

Appendix D continued from previous page.

Title of Document	Purpose	Located in Files of	
		Investigator/ Institution	Sponsor
Curricula vitae and/or other relevant documents demonstrating qualifications of investigator(s) and sub-investigator(s)	To document qualifications and eligibility to conduct trial and/or provide medical supervision of subjects.	X	X
Normal value(s)/range(s) for medical/laboratory/ technical procedure(s) and/or test(s) included in the protocol	To document normal values and/or ranges of the tests.	X	X
Medical/laboratory/technical procedures/tests	To document competence of facility to perform required test(s), and support reliability of results.	X (where required)	X

Appendix D continued on next page.

Appendix D continued from previous page.

Title of Document	Purpose	Located in Files of	
		Investigator/ Institution	Sponsor
• Certification; or • Accreditation; or • Established quality control and/or external quality assessment; or • Other validation (where required)			
Sample of label(s) attached to investigational product container(s)	To document compliance with applicable labelling regulations and appropriateness of instructions provided to the subjects.		X
Instructions for handling of investigational product(s) and trial-related materials (if not included in protocol or Investigator's Brochure)	To document instructions needed to ensure proper storage, packaging, dispensing, and disposition of investigational products and trial-related materials.	X	X

Appendix D continued on next page.

Appendix D continued from previous page.

Title of Document	Purpose	Located in Files of	
		Investigator/ Institution	Sponsor
Shipping records for investigational product(s) and trial-related materials	To document shipment dates, lot numbers/serial numbers and method of shipment of investigational product(s) and trial-related materials. Allows tracking of product, review of shipping conditions, and accountability.	X	X
Testing or conformity certificate(s)	To document evidence of compliance with directives or standards (product specific or horizontal) to demonstrate device safety. May be certificates issued by third parties.		X
Decoding procedures for blinded trials	To document how, in case of an emergency, identity of blinded investigational product can be revealed without breaking the blind for the remaining subjects' treatment.	X	X (third party if applicable)

Appendix D continued on next page.

Appendix D continued from previous page.

Title of Document	Purpose	Located in Files of	
		Investigator/ Institution	Sponsor
Master randomisation list	To document method for randomisation of trial population.		X (third party if applicable)
Pre-investigational monitoring report	To document that the site is suitable for the trial (may be combined with study initiation report).		X
Study initiation monitoring report	To document that trial procedures were reviewed with the investigator and the investigator's trial staff (may be combined with pre-investigational report).	X	X

Appendix D continued on next page.

Appendix D continued from previous page.

During the Clinical Trial

The following should be added to the files during the trial as evidence that all relevant information remains documented as it becomes available.

Title of Document	Purpose	Located in Files of	
		Investigator/ Institution	Sponsor
Investigator's Brochure updates	To document that investigator is informed in a timely manner of relevant information as it becomes available.	X	X
Any revision to: • Protocol/amendment(s) and CRF • Informed consent form • Any other written information provided to subjects about device	To document revisions of these trial-related documents that take effect during trial.	X	X

Appendix D continued on next page.

Appendix D continued from previous page.

Title of Document	Purpose	Located in Files of	
		Investigator/ Institution	Sponsor
• Procedure for subject recruitment (if used)			
Dated, documented approval/favourable opinion of Institutional Review Board (IRB)/ Ethics Committee of the following: • Protocol amendment(s) • Revision(s) of: • Informed consent form • Any other written information to be provided to the subject	To document that the amendment(s) and/or revision(s) have been subject to IRB/Ethics Committee review and were given approval/favourable opinion. To identify the version number and date of the document(s).	X	X

Appendix D continued on next page.

Appendix D continued from previous page.

Title of Document	Purpose	Located in Files of	
		Investigator/ Institution	Sponsor
• Procedure for subject recruitment • Any other documents given approval/favourable opinion • Continuing review of trial (where required)			
Approval/notification of trial from appropriate Regulatory authorities (if required)	To document appropriate regulatory approval/notification obtained prior to initiation of the trial. For commercialised products, copy of certificates of CE-marking from NB or similar approvals.	X (where required)	X

Appendix D continued on next page.

Appendix D continued from previous page.

Title of Document	Purpose	Located in Files of	
		Investigator/ Institution	Sponsor
Curriculum vitae for new investigator(s) and/or sub-investigator(s)	To document qualifications and eligibility to conduct trial and/or provided medical supervision of subjects.	X	X
Updates to normal value(s)/range(s) for medical/laboratory/technical procedure(s)/test(s) included in the protocol	To document normal values and ranges that were revised during the trial.	X	X
Updates of medical/ laboratory/technical procedures/tests • Certification; or • Accreditation or	To document that tests remain adequate throughout the trial period.	X (where required)	X

Appendix D continued on next page.

Appendix D continued from previous page.

Title of Document	Purpose	Located in Files of	
		Investigator/ Institution	Sponsor
• Established quality control and/or external quality assessment or • Other validation (where required)			
Documentation of investigational product(s) and trial-related materials shipment	Document shipment dates, lot/serial numbers and method of shipment of investigational product(s) and trial-related materials. Allows traceability of product lot, review of shipping conditions, and accountability.	X	X
Testing or Conformity Certificate(s)	To document evidence of compliance with directives or standards (product specific or horizontal) to demonstrate safety. May be certificates issued by third parties.		X
Monitoring visit reports	To document site visits by, and findings of, the monitor.		X

Appendix D continued on next page.

Appendix D continued from previous page.

| | | Located in Files of | |
| | | Investigator/ Institution | Sponsor |
Title of Document	**Purpose**		
Relevant communications other than site visits • Letters • Meeting notes • Notes of telephone calls	To document any agreements or significant discussions regarding trial administration, protocol violations, trial conduct, adverse event (AE) reporting.	X	X
Signed informed consent forms	To document that consent is obtained in accordance with GCP and protocol and dated prior to participation of each subject in trial. Also to document direct access permission.	X	
Source documents	To document the existence of the subject and substantiate integrity of trial data collected. To include original documents related to the trial, to medical treatment, and history of subject.	X	

Appendix D continued on next page.

Appendix D continued from previous page.

Title of Document	Purpose	Located in Files of	
		Investigator/ Institution	Sponsor
Signed, dated, and completed Case Report Forms (CRF)	To document that the investigator or authorised member of the investigator's staff confirms the observations recorded.	X (copy)	X (original)
Documentation of CRF corrections	To document all changes/additions or corrections made to CRF after initial data were recorded.	X (copy)	X (original)
Notification by originating investigator to sponsor of reportable adverse events	Notification by originating investigator to sponsor of reportable adverse events.	X	X

Appendix D continued on next page.

Appendix D continued from previous page.

Title of Document	Purpose	Located in Files of	
		Investigator/ Institution	Sponsor
Notification by sponsor and/or investigator, where applicable, to regulatory authority(ies) and IRB(s)/ Ethics Committees of unexpected adverse events	Notification by sponsor and/or investigator, where applicable, to regulatory authorities and IRB(s)/EC(s) of adverse events.	X (where required)	X
Interim or annual reports to IRB/Ethics Committee and authority(ies) (if required)	Notification by sponsor to investigators of safety information in accordance with 5.16.2. Interim or annual reports provided to IRB/Ethics Committee.	X	X (where required)

Appendix D continued on next page.

Appendix D continued from previous page.

Title of Document	Purpose	Located in Files of — Investigator/Institution	Located in Files of — Sponsor
Subject screening log	To document identification of subjects who entered pre-trial screening.	X	X (where required)
Subject identification code list	To document that investigator/institution keeps a confidential list of names of all subjects allocated to trial numbers upon their enrolment in the trial. Allows investigator/institution to identify any subject.	X	
Subject enrollment log	To document chronological enrollment of subjects by trial number.	X	
Investigational products accountability at the site	To document quantity and number of investigational product(s) used according to the CIP.	X	X

Appendix D continued on next page.

Appendix D continued from previous page.

Title of Document	Purpose	Located in Files of	
		Investigator/ Institution	Sponsor
Signature sheet	To document signatures and initials of all persons authorised to make entries and/or corrections on CRFs.	X	X
As appropriate, record of retained body fluids/tissue samples	To document location and identification of retained samples if assays need to be repeated.	X	X

Appendix D continued on next page.

Appendix D continued from previous page.

After Completion or Termination of the Trial

After completion or termination of the trial, the following documents should be retained in the file together with all documents mentioned in previous sections of this Appendix.

Title of Document	Purpose	Located in Files of	
		Investigator/ Institution	Sponsor
Investigational product(s) accountability at site	To document that the investigational product(s) have been used according to the protocol. To document the final accounting of devices received at the site, used, and returned by site to sponsor.	X	X
Documentation of investigational product destruction, if appropriate	To document destruction of unused devices by sponsor.	X	X

Appendix D continued on next page.

Appendix D continued from previous page.

Title of Document	Purpose	Located in Files of	
		Investigator/ Institution	Sponsor
Completed subject identification code list	To permit identification of all subjects enrolled in the trial in case follow-up is required. List should be kept confidential and for agreed upon time.		
Audit certificate/report	To document that audit was performed.	X	
Final trial close-out monitoring report	To document that all activities required for trial close-out are completed, and copies of essential documents are held in the appropriate files.	X	X
Treatment allocation and decoding documentation	Returned to sponsor to document any decoding that may have occurred.		X

Appendix D continued on next page.

Appendix D continued from previous page.

Title of Document	Purpose	Located in Files of	
		Investigator/Institution	Sponsor
Investigator's report to EC/IRB	To document completion summary of the trial and its progress by the investigators for submission to the EC/IRB.	X	
Clinical study report	To document results and interpretation of data from the trial.	X (if applicable)	X

23

Adverse Event Processing During Clinical Trials

Kenneth R. Michael

KRM & Associates
San Diego, CA, USA

Marja G. de Jong

Medidas
Wijk bij Duurstede, The Netherlands

Under the medical device requirements, reporting of adverse events is part of the *post-marketing surveillance* system that a company is obligated to implement. In principal, adverse event or vigilance reporting is to be done for all products, regardless of whether the products are

- commercialised;

- under clinical evaluation; or

- classified under the Medical Device Directive (MDD) (93/42/EEC) as Class I, IIa, IIb, or III.

During a clinical trial, claims regarding the intended use of the device are established. By the time a clinical trial commences, not all of the experiences concerning the device's performance in normal use in humans may have been defined. All known risks, as well as benefits, however, should have been established prior to the clinical trial. The criteria for moving into a clinical trial comes from the *in-vivo* studies or pre-clinical data. Some of these trial results can contribute to the product; others may contribute to the safe use of the device.

All adverse events are to be recorded and investigated, and corrective actions are to be implemented. Only reporting of the clinical problem to the regulatory authorities is required; no reporting of potential treatments is necessary. Trending and tracking of the adverse events in relation to the device and to its predecessors overall is required in addition to recording and reporting.

Under a clinical trial, the results of the clinical investigation, as well as the corresponding corrective or preventive actions, may have a direct impact on the clinical protocol and the process of the trial itself. The importance of any adverse event can become critical in determining the actual performance and characteristics of the device.

This chapter will examine the process of adverse event reporting during a clinical trial. All the aspects referred to in this chapter are also applicable to adverse event reporting after the device has been commercialised. Although there are some minute differences in some countries, the basic elements of recording, handling, reporting, investigating, and closing of adverse events are similar.

LEGISLATION/REQUIREMENTS

According to the regulations governing medical devices[1], adverse events are reportable events if they meet any of the following:

---

1. The list is a compilation of the reportable events found under the European medical device legislations (MDD and AIMD) as well as FDA requirements. Not all of the elements listed may apply to all non-European countries.

- death;

- life-threatening so medical intervention becomes necessary;

- causes or contributes to serious injury;

- technical or medical reason in relation to the characteristics or performance of the device which may lead to a recall of the product; or

- involves a malfunction (or near event) of the product.

In brief, adverse events can be categorised as any event that is contrary to the indications, performance, and safe use of the device. Adverse events are **unexpected** or **unanticipated events** that are not part of the original product's risk analysis nor part of the acceptable risks under which the device can still be safely used.

Once a manufacturer[2] or clinical trial sponsor has determined that an adverse event has taken place, there is a time limit in which to report the adverse event to the relevant regulatory authorities. In the case of a clinical trial, reporting of these events is to appropriate regulatory authorities in the countries where the clinical trial sites are located. For multi-national multi-centre trials, this would mean reporting the events to **each** national regulatory authority where the trial site is located. In addition, specifically for the United States, if the clinical trial sponsor is an American-based or American-owned company, the FDA requires that the event be reported to its organisation as well.

For Europe, reporting times are a little more convoluted. CE-marked devices which are the subject of clinical investigation for reasons other than those mentioned in chapter 1, fall within the scope of the vigilance reporting guidelines defined in the MDD as amplified by the guidance on vigilance (MEDDEV 12.1/1 formerly 3/93). Vigilance reporting requirements apply to devices, both CE-marked and non-CE-marked devices where

---

2. A manufacturer and a clinical trial sponsor are considered, for the sake of this chapter, to be identical. Throughout this chapter, the term *clinical trial sponsor* will be used to designate any individual or organisation initiating a clinical trial on a device, whether the device is a new technology or an existing one.

such events lead to corrective actions relevant to CE-marked devices. Under the reporting requirements, time of filing an initial report is dependent on whether the anomaly is an adverse event or near event (i.e., malfunction). Adverse events are to be reported within ten days; near events, within thirty days.

Vigilance reporting, in principle, does not apply to devices under clinical investigation. Yet, *adverse events* which arise from clinical investigation on devices *do* require reporting.

Annex X clearly states 'all adverse events as these specified [previously] must be fully recorded and notified to the CA. Clinical investigational adverse events are to be reported to the authorities (i.e., EC and CAs) of the country where the clinical trial site is located. Furthermore, a number of the GCP documents referenced in chapter 1 require reporting of adverse events that arise during a clinical trial. For multi-centre trials, this would mean reporting the events to *each* national regulatory authority of the country *where the trial sites are located.* Reporting the adverse event is clear; the time of reporting is still an issue. The reader should inquire of the CAs of the Member States where the trial sites are located how reporting of clinical investigational adverse events is to be handled. Should the CAs not have a clearly defined requirement, utilising the reporting times stipulated in MEDDEV 12.1/1 should be followed for all reportable adverse events.

Under the FDA requirements (1996), deaths, serious injuries, and malfunctions are required to be reported within 30 days. Any device-related events that require immediate action to protect public health are required to be reported within 5 days.

Any adverse event that does not have a clinical basis does not necessarily need to be reported to the regulatory authorities. Nonetheless, these are to be included in the overall trend analysis of the complaints and adverse events.

Implementation of adverse event handling has various steps. Initially, there is a report from the user, followed by an investigation; determination and implementation of corrective actions; then closure and reporting of the adverse event to the appropriate regulatory authorities. The following sections will examine these aspects in some detail.

HOW THE ADVERSE EVENT IS TO BE REPORTED TO THE SPONSOR

Unexpected or unanticipated adverse events arising during clinical investigation are to be reported by the clinician to the monitor and/or the sponsor immediately (i.e., within 24 hours of occurrence). Adverse events that occur during a clinical trial are mostly **un**expected or **un**anticipated events[3]. These **must** be *reported* and analysed in a trend database. Should the event be anticipated (i.e., part of the acceptable risks under which the device can be used safely, or part of the risk analysis of the device itself), then these may not be reportable events to the regulatory authorities; they do, however, require trending.

The clinician is the individual responsible for writing the initial adverse event report. The following aspects need to be included:

- description of the event that occurred;

- lot or serial number of the product involved;

- name of the product as described in the accompanying proposed labelling;

- condition of the patient at the time of occurrence and present status at time of reporting;

- medical intervention given to the patient as a remedy to the adverse event or other follow-up activities[4];

- identification number or initials of the patient along with the patient's birth date. Do not use numerical designations for the month since this may be confusing. Write out the month in full or abbreviate.

This report may be a statement written by the clinician or a form that is completed by the clinician. Appendix C provides a sample of a reporting format which can be used for adverse event reporting during a clinical trial. This sample is also useful in the reporting of complaints (including adverse events) of

---

3. See definition previously stated (page 483).

4. Clinical treatment is not reportable of itself.

commercialised products. If the clinical trial sponsor decides to use an adverse event reporting form, this form should be included in the protocol; a number of separate copies could be provided to the clinician and his/her staff.

The initial adverse event report is sent by **fax or other communication medium** to the Regulatory Affairs contact person(s) within the sponsor's organisation. The names, addresses, and telephone and fax numbers of these contact persons are required to be included in the protocol. Since these contact persons must be reachable at all hours, the inclusion of telephone and fax numbers during and after office hours is essential.

TIP	Without a contact person, the adverse event commonly disappears into a black hole! *Designate one knowledgeable contact person for any study.* It is crucial to have an SOP within the company defining the contact person and his/her responsibilities.

The information that the doctor can provide to the sponsor at the time of the adverse event, as well as during the sponsor's investigation, is guided by the privacy laws in the countries where the trials are conducted. For U.S.-based as well as European-based companies, it is important to understand that privacy laws protect the patient as well as any information about the patient. In the United States, some of this privacy information is found within the Investigational Device Exemption (IDE) requirements of the Code of Federal Regulations (CFR). In the European Union, privacy is part of GCP supplemented by directive 95/46/EEC,[5] and national privacy laws at the member state level.

Doctors worldwide are bound by their oath as medical practitioners to treat all information about their patients as confidential. Even if desired by the clinical trial sponsor, it may not always be possible to obtain, for instance, written permission from the patient in the case of explantation of an implanted device. These elements are to be taken into consideration during the design of the protocol and the informed consent. The minimum information that can be provided to the

---

5. European Directive 95/46/EEC on the protection of individuals with regard to processing of personal data on the movement of that data.

sponsor/manufacturer in case of device explanation can be the patient identification number or the patient's initials and his/her date of birth.

| TIP | Understanding the privacy or patient data protection requirements in the countries where the clinical trial is being conducted is crucial. Aspects of these requirements are to be included in the protocol, informed consent, and the criteria for handling and reporting of adverse events. |

INVESTIGATION OF THE
ADVERSE EVENT BY THE SPONSOR

Once the report is received by the regulatory contact person, the adverse event is subject to investigation. The accuracy of the initial information is determined by the technical evaluation of the product by the user.

To make the investigation process easier, it is advisable to develop a list of questions or define a plan for the investigation of the reported adverse event. This should define the strategy for handling the adverse event. The questions need to be defined in such a manner that they are easily understood by the clinicians and the clinical trial monitor.

| TIP | Focus on and define the *adverse event* prior to starting the investigation! Create a standard format for reporting adverse events (see sample form in Appendix C). |

Use of clinical trial monitors to go into the clinical site to ask the questions is a suitable manner by which to obtain further valid information. Regulatory officials at the sponsor's site can also enter the clinic and ask questions. Gather the information factually, without any opinions and personal viewpoints.

| TIP | STOP what you are doing! Look directly at the individual providing the information. Listen not only to the words but also be aware of the emotion, concern, and buzz words being used. KEEP QUIET and do not be DEFENSIVE. Retrieve information; do not give information this time! |

When composing the questions, always consider to whom you are directing the questions. Typical questions include the following.

- What is the condition of the patient at the time of the incident, including the diagnosis at the time of the incident as well as any reaction to potential treatments?

- What is the condition of patient after the incident?

- Has the incident been remedied by medical treatment?

- What type of treatment was given to the patient to alleviate or remedy the incident?

- What medications were given?

- Who was present at the time of the incident?

Try to obtain access to the medical records **to read** the treatment and description **recorded** by the attending physician at the time of the incident. In some countries, this may not be permissible and is subject to privacy laws or data protection laws. In such cases, a written medical statement from the treating physician is advisable, or else an accurate recording of verbal statements may be made by the doctors or nursing staff. In the latter case, a counter-signature by the informing nurse or medical practitioner may be advisable. This is a method which maintains compliance with the privacy laws while ensuring the accuracy of information being provided.

NOTE	Remember, the objective is to obtain all of the known facts in an objective and non-belligerent, non-defensive manner.

It is important to acquire all of the data and information about the incident during the investigation. This could take from several hours up to several days.

PROCESSING THE INVESTIGATION'S FINDINGS

Once this information has been gathered, it is then necessary to perform a risk analysis. This is done by reviewing the product's original risk analysis as well as any risks associated with the incident itself. Use of appropriate methodology such as Failure Mode and Effects Analysis (FMEA), Fault Tree Analysis (FTA), or general medical device risk analysis[6] is important. Whichever methodology is employed, it is important to ask the following questions.

- Does the event fall within the scope of the risks set for the product in the labelling (i.e., warnings, cautions)?

- Does the event indicate an acceptable risk?

- What are the trend results from the pre-clinical data?

- What are the trends of products that are substantially equivalent to the product under clinical investigation? (This question will be harder to answer if the product is a new technology.)

The results of the risk analysis need to be mitigated and a corrective action plan established.

 NOTE Analyse the information by using the company's risk analysis procedure. This procedure needs to be based on appropriate methodologies pertaining to risk analysis (e.g., EN 1441[7], IEC 812[8], IEC 56410[9], ISO/CD 14971[10], IEC 1025[11]). Determine the strategy for mitigating and correcting the event.

In cases of implanted products, these could be explanted if the clinician finds that this may remedy the patient's symptoms. Such explanted devices need to be returned to the manufacturer

---

6. EN 1441, Medical Devices: Risk Analysis.

7. This standard is a harmonised standard to meet the requirements of the European medical device legislation (MDD 93/42/EEC and AIMD 90/385/EEC). To apply it, also review the FMEA and FTA requirements defined in the other risk assessment standards.

8. IEC 812, FMEA: Analysis Techniques for System Reliability: Procedures for Failure Modes and Effects Analysis.

9. IEC 56410, Risk Analysis of Electrical Systems—Application Guidance.

10. ISO/CD 14971, Risk Assessment.

11. IEC 1025, Fault Tree Analysis.

for product evaluation. This evaluation may include destructive product testing.

The objective of the investigation, both in interviewing the medical staff as well as product evaluation, is to determine the root cause of the incident. Appendix D contains an overview of the **root cause** analysis of reported problems.

CORRECTIVE ACTIONS

Once the cause has been determined through investigation, the corrective action needs to be identified. In defining the corrective action and its implementation, there are two aspects that need to be considered.

1. Is the product under clinical investigation already a commercialised product?

2. Is the product a non-commercialised product?

If a product is already commercialised or the product is a new technology under clinical investigation, this will determine how its corrective actions are implemented and processed.

Commercialised Product

For commercialised products, there is a feedback mechanism back to the design and production phases of the device. The flowchart in Appendix A illustrates a process for the handling of these types of events and the ultimate implementation of corrective actions. From this flowchart, it can be concluded that adverse events for commercialised products used in a clinical investigation require to be brought into the overall post-marketing surveillance/feedback mechanism established by the company. In practice, this means that the information obtained during the investigation of the adverse event is to be assessed against the design and production aspects of the product cycle.

Major corrective actions could be required or minor corrective actions, such as labelling changes, could result. Attention also needs to be paid to the clinical protocol used when such a commercialised device is the subject of a clinical trial.

Amendments as well as Ethics Committee or Institutional Review Board (IRB) approval is also required. In this situation, the reader should be aware that one of the corrective actions that could occur is a product recall.

Non-Commercialised Product

Should a new-technology product be involved in a clinical trial investigation, the corrective or preventive actions could impact both the product and/or the clinical investigation. All processes pertaining to production and design are involved, as well as modifications to the clinical trial process. Appendix B is a flowchart that outlines the process for handling these types of events and the ultimate implementation of corrective actions.

For both situations, the corrective action **must** be detailed. A timeline needs to be defined and established. The following questions should be answered and considered when planning the strategy for the effective implementation of corrective actions.

- When is the corrective action to be completed?

- By whom is the corrective action to be implemented?

- Does the corrective action include a root cause determination?

- What is the corrective action to be undertaken? Does it include a review of the risk analysis?

- What is the impact of the corrective action on the device, clinical investigation, and so on?

Specifically, for product under trial, consider:

- changes to product labelling, software, hardware, and so on;

- changes to the protocol;

- a request of change approval by the IRB or Ethics Committee; and

- information, including the training of clinicians, to ensure corrective action is understood and implemented by them during the remainder of the trial.

Other aspects that affect the manufacturer's processes should also be considered, for example, design specification, process validation, manufacturing processes, and materials.

Basically, any corrective action may include a change to the labelling, software, hardware, design, manufacturing, materials, and/or protocol. The root cause of the adverse event should determine the type of corrective action to be implemented. This could be the most difficult item to determine since it requires an unbiased, knowledgeable clinical viewpoint. It is worth considering using an impartial and unbiased scientific approach, for example, a forensic pathologist.

| TIP | Define the corrective action based on the impartial and objective, professionally expressed information! Use an unbiased, independent/third party that has both the expertise and experience to assist in understanding and positioning the corrective action, where and when possible. |

NOTIFICATION OF CORRECTIVE ACTIONS TO RELEVANT PARTIES

Once corrective action is determined, it is to be stipulated in a report or a type of advisory notice (not unsimilar to the one required for recall procedures). This document should include

- the background of the adverse event;
- the lot and/or serial numbers of the product;
- the description of the product;
- the initial investigation findings; and
- corrective actions.

This report or notification is sent to appropriate regulatory authorities as well as to the clinicians. Send such notifications via registered post or courier to have evidence of delivery and receipt.

If the corrective action is a temporary one, amendments or supplements to the protocol can be made **if the change or**

adverse event has been investigated and determined not to affect safety. Protocols, being controlled documents, are subject to the change-control procedures within the quality system of the clinical trial sponsor. All changes or amendments need to be documented and authorised accordingly. This applies to whether the corrective action is temporary or permanent[12].

NOTE	Reports or notifications should be short and contain all relevant information. In some countries, however, there may be a reporting format to be used as required by regulatory authorities. Using this as a basis for the notification may be helpful. Also, reporting could be required to be done in the national language of the country where the adverse event occurred.

Clinicians should be verbally informed about the corrective actions, along with the date when the corrective action is to be implemented. An investigator's meeting may be advisable to ensure that all of the investigators involved with the trial understand the incident and the ensuing corrective actions. Obtaining a signed commitment from the investigators is an important mechanism for implementation control. This can be done through implementing the corrective action through a protocol amendment that the clinicians are required to approve and sign. Amendments are thereby easier to present to the IRB/Ethics Committee as part of the protocol update.

TIP	Meet with the clinicians to discuss the corrective actions and have them provide written affirmation of their understanding and their commitment to the implementation. Written confirmation/signatory must be obtained from each clinician, without fail! Having them sign the amendment to the protocol is an effective and easy way of obtaining this commitment.

FOLLOW-UP

It is not sufficient to only determine and notify corrective actions. Follow-up on the effective implementation of corrective actions is essential. It ensures timely and effective implementation at all sites of the proported changes; it also ensures

---

12. 21 CFR 803.

that the trial data remain valid and accurate. These activities are often forgotten because clinicians, as well as clinical trial sponsors, do not understand how to undertake follow-up actions.

It is important to send an experienced clinical trial monitor to the clinic. Plan the follow-up activity in the shortest possible time frame after notification of the corrective action. Furthermore, ensure that all of the clinicians involved in the clinical trial effectively implement the corrective actions at the same time. This is a "must". Written commitment as previously stated can function as an added incentive.

It cannot be stressed sufficiently by the authors that follow-up is done to ensure that the corrective action is **consistent and effectively** implemented throughout the clinical investigation and at all sites. Corrective actions need to be fully

TIP	Send the monitor into the clinic for follow-up with a checklist and document defining who did what, when it was done, and where it was done.

implemented. Regulatory bodies will look for a 90 percentile range for effectivity of the corrective action as a basis of criteria for assessment.

TIP	Do not forget to monitor the manufacturing site as well, since in corrective actions there is always more than one party responsible for the corrective action.

CLOSURE OF AN ADVERSE EVENT

The follow-up needs to be continued for a specific period of time in order to achieve 90 percentile effectivity. Once this is achieved, a final report may be written concerning the event. This report is forwarded to all appropriate regulatory authorities where trial sites are located. An executive summary is to be retained in the product's adverse event file as well as in the clinician's and the investigational dossier.

Once the event has been closed, the objective has not yet been completed. The information needs to be fed into a post-marketing surveillance system that gives access to the tracking of devices and trending of device-related issues. Part of the closure of the adverse event is the trending and archiving of the incident.

All adverse events that occur during a clinical trial are to be trended. This applies to all products involved in the clinical trial, whether the investigational device is already commercially available or not (in the case of a new technology).

Trends are basically statistical analyses of the performance of the device viewed against its original product specifications. Any anomalies that emerge during the trends are to be separately analysed. Trend analysis is multi-functional: It can provide information that can be indicative of a potential safety problem with the product; it can also point to preventive or corrective actions that are required.

Many companies have queries concerning the archiving of adverse events. Maintaining a *separate* adverse event file is recommended. Under the FDA requirements, a separate medical device reporting (MDR) file is required[13]. It is not advised to retain all adverse events together with any other product complaints. Specific to clinical trials, there must be an overall audit trail of adverse events in relation to the clinical site and the specific clinical trial. Traceability back to the actual clinical trial, hospital site, and so on, therefore, needs to be established.

CONCLUSION

The reporting of adverse events, along with their investigation, is an important element in the overall feedback mechanism concerning a device's actual performance in a clinical setting. Whether the product is one under clinical trial or one that has already been on the market for many years, the process defined in this chapter can help to establish a mechanism for the reporting of, investigation of, and determination of corrective actions and follow-up, and closure of adverse events and product complaints in general.

---

13. 21 CFR 803.

The complaint process, whether adverse events or not, follows the basic elements. Hereunder referenced:

- Focus on the adverse advent itself.

- Investigate to obtain all of the facts.

- Analyse the data.

- Determine the corrective or preventive action.

- Execute or implement the corrective action with corrective action follow-up and closure of the complaint or event.

In the execution phase, reporting of the incident to the appropriate authorities and the archiving of the findings and results also need to be done.

Although the emphasis of this chapter is on the clinical trial, the contents and processes defined herein are also applicable to devices that have already been commercialised. It is, however, important to remember that, under the various reporting guidelines and regulations in today's medical device environment, adverse events that are considered to be **unexpected or unanticipated** events—i.e., those that are not part of the device's risk analysis nor part of the device's acceptable risks—should be reported.

Any information concerning the performance of the device as defined in the manufacturer's literature (i.e., instructions for use, labelling) need not be reported to the regulatory authorities. Periodically reviewing the product's risk analysis against the complaints reported concerning the device is an important element in maintaining quality and safe products in the market and ensuring continued compliance with the product's intended purpose and specifications.

APPENDIX A. FLOWCHART: ADVERSE EVENT HANDLING OF COMMERCIALISED PRODUCTS

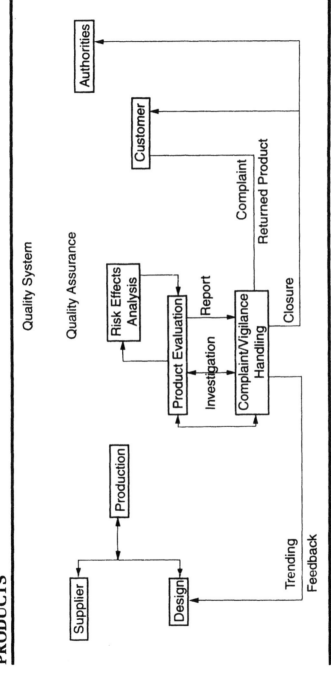

APPENDIX B. FLOWCHART: CLINICAL TRIAL ADVERSE EVENTS

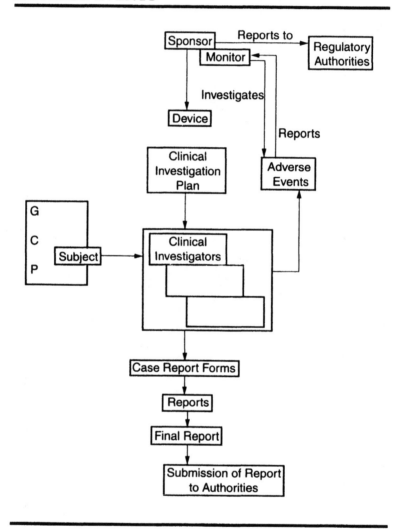

APPENDIX C. SAMPLE FORMAT FOR REPORTING ADVERSE EVENTS

General Reporting Information:

Name of person reporting adverse event:

_____

Name and address of institution:

_____

Today's date: _____

Name of person taking information about adverse event:

_____

Product Information:

Name of device as defined in its proposed labelling:

_____

Catalogue numbers: _____

Lot numbers of device: _____

Incident Information:

Date adverse event occurred:_____

Date adverse event first reported to
    sponsor/manufacturer:_____

Description of adverse event:_____

Patient Information:

Sex and age of patient: _____

Condition of patient at time of adverse event occurring:

_____

Description of medical treatment provided to patient to
remedy adverse event:

_____

Name of person who prescribed medical intervention:

_____

Condition of patient after intervention:

_____

APPENDIX D. ROOT CAUSE ANALYSIS OF REPORTED PROBLEMS

- Determine known facts and separate from conjecture.

- Collect preserved evidence.

- Determine if the cause was internal (such as in production); external (such as in troubleshooting); or a defective supplier item.

- If possible, have the supplier determine the root cause for its items.

- Determine recommendations for corrective action based on the detectability of the problem at the earliest stage possible:
  - supplier
  - inspection
  - assembly
  - test
  - installation
  - acceptance
  - preventive maintenance

Note that root cause analysis and resultant corrective action may be incomplete if the information is inaccurate or insufficient.

Section V

How to Analyse and Report Clinical Data

24

Statistics

Henriët E. Nienhuis
Parexel/MIRAI BV
Amsterdam, The Netherlands

The role of the statistician in a clinical study is still underestimated. In most cases, the statistician is involved in a study when there is a question of how large a sample size is needed. In fact, the answer to this question has already been determined, but only a statistical justification is needed.

This is very unfortunate, as the statistician is not only able to produce an adequate sample size estimation, but he/she can also provide some very valuable information with regard to the conduct of the study. Again, as with data management, it is very important to have statistical input from the beginning of a clinical study. The purpose of this book is not to give a complete overview on the statistical methods available for analysing the data of a clinical study on a medical device; the information would be too technical; and, it is nearly impossible to give enough details to allow the reader to perform the statistical analysis without further background. Therefore, the reader is referred to a statistician, who will be able to analyse the data properly. However, for reasons of completeness, the role of statistics and some basic issues of importance with regard to studies within the world of medical devices will be given.

GENERAL ITEMS OF IMPORTANCE

With respect to the most frequently asked question of the statistician (i.e., 'How large a sample do I need?'), a couple of items are of importance. First of all, the objective of the study is important. For medical devices, the design of a study will most often be comparative, or observational. In a comparative study, there will be a 'gold standard', which is the currently accepted device. The performance of the new device is then compared to this standard, and the size of the sample needed depends on the acceptable difference between the two devices from a clinical point of view. Furthermore, it is important to know what level of significance one is interested in (usually 5 percent) and what the power of the study should be (80–90 percent is fairly good). Decreasing the power of the study or increasing the level of significance results in a smaller sample size. However, the maximum sample size may be restricted due to economic reasons or practical implications. This is an undesirable situation, as a reduction in the number of patients to be included may affect the final outcome of the study. The costs of a clinical study should, therefore, never limit the number of patients to be included.

In case practical implications are the reason for using a non-optimal sample size, one should thoroughly analyse whether solutions can be found for this aspect. Thus, statistical input can be very valuable. In all cases, it is very useful to perform a power calculation prior to the study, as this can help establish whether the performance of the study is still sensible.

Furthermore, the statistician should be involved in developing the case report form (CRF). This ensures that the data to be collected are collected in such a way that the questions to be solved from the study can be answered during the statistical analysis.

 TIP To complete a statistical report of the results of the study very soon after the database has been cleaned, the statistician should do the programming part of the analysis during the data collection process.

In the ideal situation, the statistician should be involved from the beginning of the study. Ideally, the manufacturer of the device, the statistician, and the clinical specialist should write the protocol together. The statistician can give input on the data to be collected and the structure of these data, whereas the manufacturer can provide the exact question(s) to be answered by the study. Last, but not least, the clinician best knows what is convenient and reasonable in practice.

During the trial, it is important to keep the statistician informed as to the quality of the data. In particular, when special cases occur, the statistician should know how these were handled.

TIP	When a list of special cases is kept during the study, it provides information to the statistician so that these cases can be handled in a standardised fashion.

After the statistical analysis of the study has been carried out, a report of the results has to be written. Two types of reports are common: the integrated report, in which the statistical results are incorporated with the clinical implications, and the statistical report, in which the results of the statistical analysis are summarised. This report needs to fulfil all criteria to satisfy GCP (i.e., in U.S., EU, etc.).

TIP	The report should be written in such a way that the statistical analysis can be repeated in exactly the same way as it was done.

It is very useful to have the report written by the statistician in collaboration with clinicians involved. This prevents the writing of a very technical report, where in clinically important items either are not covered or are only marginally covered; a balanced report is critical.

The continuous flow of data entry, data validation, and query generation necessitates an organised and transparent way of collaboration between the persons involved. In general, five things should be kept in mind.

1. Data are monitored → data entry department

    *   Administer incoming CRFs.

| TIP | A presentation of the results to *all* persons involved in the trial is usually appreciated, even when it is presented six months after the study has been completed. |

- Inform data entry department and data management about what came in.

- Set timelines: when to enter, validate, and query data.

- Communicate timelines to monitor with respect to subsequent visits to be performed.

- Endpoint: Hand over responsibility to the data entry department.

2. Data entry and verification

- Inform data management about progress of data entry and verification.

- Report on the process with regard to timelines and agreements.

- Endpoint: Hand over clean data (double entered and corrected) to data manager.

3. Data validation and query generation

- Inform monitor on progress of data validation and query generation.

- Send queries to investigator and monitor.

- Endpoint: Send queries to investigator.

4. Query resolution

- Help investigator resolve queries.

- Make sure data corrected are clear for data entry personnel.

- Return CRFs to data entry department.

5. Data management endpoint: Deliver a clean database to the statistical department.

25

Reporting of Data

Elizabeth A. M. van der Linden

Parexel/MIRAI BV
Amsterdam, The Netherlands

The reporting of data from a clinical study is an important process. The data gives the medical community evidence that use of a new device or treatment is justified. The data is to be reported in a clinical study report, which is the culmination of results of the study. Per the directives, a final report is required; for sake of clarification, however, the term preferred is Clinical Study Report (CSR). The data can be reported either as a

1. published article; or

2. as a clinical study report.

There are several commonalities between writing an article and writing a clinical study report. Issues to be covered in the report are

*   why the study was conducted (hypotheses);

*   how the study was conducted (methodology);

*   what results were obtained (statistical analysis and conclusions); and

- what the results mean in the progress of medical practice (discussion and recommendation for further studies).

The article or report is to be complete and free from ambiguity; the underlying data must be true; and the report needs to have a logical structure and be easy to read.

There are, however, also a number of characteristics specific to writing an article compared to the clinical study report. For a medical device company, the clinical study report is the most critical document for demonstrating, in a disciplined and valid scientific manner, that clinical evidence has been obtained to show compliance with the requirements. In this chapter, we will emphasise the clinical study report. For articles, refer to Table 25.1.

PLANNING A CLINICAL STUDY REPORT

The clinical report is the end product of a range of processes inherent to clinical studies. The medical writer is often confronted with two problem areas resulting from the study process.

1. Limited time: Planning a clinical study, the timeline often is in general too optimistic; there may be delays in Institutional Review Board (IRB)/Ethics Committee approvals, patient accrual, and material problems.

2. Errors creep into the design, conduct, and analyses of the study. A study that is not well designed and conducted will never lead to an adequate study report.

To prevent or minimise these problems, the medical writer should be involved in the study from beginning to end.

The following checklist can be used to reduce corrections during the clinical report writing process, to adhere to regulatory requirements, and to minimise costs.

- Check the protocol design and statistical analyses plan.
- Check the protocol amendment procedures.
- Check if deviations from the planned protocol have occurred.

Table 25.1. Elements of a Scientific Article

Title	
Authors	names, scientific degree in a separate part of the article • institutions, departments, addresses of authors • principal author for reprint requests and correspondence
Abstract	short summary of background, methods, results, and conclusions
Introduction	short overview of the reason why the study was conducted
Methods	• patients who are eligible • treatment plan • assessments • statistical analyses
Results	• patients: number of patients entered in which period • compliance of treatment • outcome findings: efficacy, safety
Discussion	• conclusions • explain methodological shortcomings • describe place of the study in relation to other relevant publications • give recommendations for further research
Acknowledgments	
References	specified by the journal
Tables and Figures	
Manuscript Format	Every journal has its own specifications for the format. Before writing, read these specifications or consult the uniform requirements for manuscripts submitted to biomedical journals: *N. Engl. J. Med.* (1991) 324:424–428; *BMJ* (1991) 302:338–341.

- Check the cleanliness of the database.
- Check the statistical analyses program.
- Check the timeliness of every activity in the clinical investigation.
- Check the quality assurance and quality control systems.

To limit time pressure, start writing the clinical study report at the beginning of the study. Follow the timelines of the study and communicate in a timely fashion to the responsible project leader what delays mean for the report writing. Provide the project leader with updated, realistic contingency plans.

THE CLINICAL STUDY TEAM AND REPORT WRITING

The final responsibility for the clinical study report rests with the project leader of the study. The project leader will delegate the operational activities to the statistician and medical writer. Statisticians and medical writers should be qualified according to Good Clinical Practice (GCP) guidelines. The statistician and medical writer give feedback to the other team members (e.g., project leader, monitors, database managers, data entry persons) during the problem-solving processes. It is useful to train the team members at the start of a study what they expect from the team in relation to their responsibility (i.e., statistical analyses and report writing).

A clinical study report should be a fully "integrated" report. It is not simply a matter of joining together separate clinical and statistical reports. The clinical and statistical description, presentation, and analyses are integrated into a single report incorporating tables and figures into the main text or at the end of the report, along with any appendices.

The statistician and medical writer have to work closely together. They form a sub-team, and it is advisable to appoint colleagues for peer review. In their ongoing training process, they need to focus on

- new guidelines emanating from regulatory bodies; and

- specific committees working on guidelines for studies with medical devices (e.g., the different classes of devices; or even a group of devices; courses for new statistical approaches; or new medical writing aspects).

GUIDELINES AND PROCEDURES

For medical devices, there is no actual current guideline for clinical study report writing. In general, the company should have, according to the GCP guidelines, written standard operating procedures (SOPs). The U.S. Food and Drug Administration (FDA) for example, has developed guidances for different types of devices (e.g., *Guidance for the Submission of Research and Marketing Applications for Interventional Cardiology Devices*). For the pharmaceutical industry, the ICH has published a guideline called *Structure and Content of Clinical Study Reports*. These two documents could give some guidance to device manufacturers regarding how to write a clinical study report. In general, the sponsor should have (in accordance with GCP) a written procedure for clinical study report format (see chapter 9).

The following elements may be included in the final report:

- title page;

- synopsis (limited to 3 pages), including numerical data;

- table of contents;

- list of abbreviations and definitions of terms;

- Ethics

  - Institutional Review Board/Ethics Committee statement of approval;

  - ethical conduct of the study, in accordance with the Declaration of Helsinki; and

  - patient information and consent (including how and when consent was given);

- investigators and study administrative structure—all parties and committees involved will have to be described;

- introduction—brief statement placing the study in the context of the development of the product;

- study objectives;
- investigation plan;

  A. Study design and plan description;

  B. Discussion of study design, including the choice of control groups;

  C. Selection of study population, including rules with respect to the withdrawal of patients from the study (i.e., stopping rules);

  D. Treatment;

  E. Efficacy and safety variables;

  F. Data quality assurance;

  G. Statistical methods and sampling,

    - statistical and analytical plan; and

    - determinations of sample size; and

  H. Changes in the conduct of the study or planned analyses;

- study subjects

  - disposition of subjects; and

  - protocol deviations;

- efficacy evaluation

  - data sets analysed;

  - demographic and other baseline characteristics;

  - measurement of treatment compliance; and

  - efficacy results and tabulations of individual patient data;

- safety evaluation;

  - extent of exposure;

  - adverse events; and

- deaths, other serious adverse events, and other significant adverse events;

- discussion and overall conclusions;

- tables, figures, and graphs referred to but not included in the text;

- reference list; and

- appendices

Another format for the clinical study report is the scientific article, as defined in Table 25.1.

STEPWISE APPROACH TO REPORT WRITING

To organise the writing of a clinical study report, one could divide the work into five phases.

Phase I: Description of the Study

The objectives of the study should cover question hypothesis or intended labelling claims for the device. The objective should be scientifically and medically relevant. According to GCP, a hypothesis is *formulated*; the objective of the study is reflected in a more metric approach (i.e., a numerical difference between subject groups concerning endpoint variables).

Investigation Plan

The *study design*, including the choice of control groups, should be described. If randomisation was not used, it is important to explain how other techniques, if any, guarded against systematic selection bias.

Selection of Study Population

Mention inclusion criteria, exclusion criteria, and the removal of subjects from therapy or assessment if applicable. Also describe screening criteria to establish external validity (i.e., subjects who are not included but were suitable for the study).

Treatment

Describe functions, application, and technical details of the investigation procedure in the study. Give characteristics of the device and specific methods used to assign a treatment to a subject. Describe the method and the justification of the control method. If applicable, describe adjunctive treatments (e.g., randomisation, blinding).

Efficacy and Safety Variables

Describe criteria used to assess the effectiveness and safety of treatment. Give a definition of the outcome variables; for many devices, there are standard definitions of success, failure, and complications (major and minor). If standard definitions are variables, show reliability, accuracy, and relevance of assessments used. Give, preferably in a flowchart, the baseline and follow-up assessment.

Data Quality Assurance

Describe the quality assurance and quality control systems that were implemented (e.g., training sessions, monitoring of investigators, operation manuals, data verification, central labs, data audits).

Statistics

The statistical analysis *plan* in the protocol and any changes made before outcome results were available should be described. Also give planned sample size methods used for sample size calculation. Also, the planned monitoring of results should be described, as well as an end point monitoring committee. If interim analyses were planned, describe the frequency and nature of these analyses and the so-called stopping rules.

The Protocol

For this phase, use the protocol. Some medical writers also author/co-author the protocol. Write the protocol in light of the clinical study report. If the protocol is adequate, this phase of the study report can be compiled at an early stage in the project.

Phase II: Presentation of Results

In phase II, any change from the originally planned conduct of the study or analyses should be described. All protocol amendments should be documented. The study population in terms of completers, noncompleters (e.g., cost to follow-up, dropouts, protocol violators) should also be summarised. Give a definition of those subjects who are included in efficacy analyses. For efficacy analyses, two types of analyses are performed:

1. Intention-to-treat analysis. All patients who entered the study are included.

2. Per-protocol analysis. Those subjects with no or minor violations are analysed separately. All patients, however, should be included in the analyses of adverse effects.

The comparability of treatment groups in terms of demographic and other baseline criteria and response to treatment should be shown. Comparability can lead to an imbalance due to the number of subjects who were lost to follow-up (e.g., patients who did not meet the effectiveness and safety criteria) and dropouts (subjects who did not meet the criteria of a completer). Analyse the effect of dropouts on the treatment. The statistical analyses should show the size of the difference between treatments, associated confidence intervals, and results of hypothesis testing. Show medical relevance of the findings after statistical analyses. Sometimes, it is necessary to re-compile statistical tables to make a clearly structured presentation to the reader.

NOTE | Before presenting the results, be sure that the definitions for each patient group are clear and that the reasons why each patient has been allocated to a special data set can be followed. Co-operate with the statistician on a regular basis during the analyses period to facilitate the report writing process.

Phase III, Introduction, Conclusions, and Discussion

Once an overview of the study results is available, start writing the conclusions. Make sure that all questions as stated in the protocol are answered. Efficiency conclusions should consider primary and secondary end points and potential exploratory analyses.

Special attention needs to be paid to serious adverse events or reactions. Risks and risk analysis should be described, and the implications of the safety evaluation for the use of the device should be discussed. Describe the benefit of the device in relation to risks and if the benefits exceed the risks. If the results of the study do not cover all of the study questions, state why this did not happen.

In the discussion, clinical relevance and the importance of results should be embedded into what is already published on the investigation product. The discussion is terminated with suggestions for further study. After the conclusions and discussion are written, it is easy to write the introduction. It is important to introduce the reader to the study and the investigational product. Describe why this study has been done in relation to the clinical development plan of the product. Explain the characteristics of the study (e.g., rationale, objectives, study population, and end points). Furthermore, give a brief description, if applicable, of agreements with authorities or guidelines and standards that were followed. In general, the introduction should be limited to one page of text.

Phase IV: Formatting

In phase IV, the report must be formatted according to the SOPs of the company. Arranging the report according to the following structure may be helpful.

- At the beginning of the report, add:
    - title page;
    - signature page;
    - synopsis;
    - table of contents;
    - ethics requirements; and
- investigators and study administrative structure.
- After conclusions, add:
    - tables, figures;
    - demographic data;
    - efficacy data;
    - safety data;
    - reference list/literature list; and
    - appendices:
        - study information;
        - patient data listings;
        - Case Report Forms (CRFs); and
        - individual patient data listing.

Phase V: Peer Review

Distribute the report to colleagues and investigators for review. If there is a deadline, start the review during the writing process. A general check to see if the writer is on the right track also helps. The report can be discussed with and reviewed by members of the medical and technical department and by investigators. In this way, many comments can be prevented from having to be acted upon at the end of the study. Everybody likes to influence the report, but there is cold comfort for all of them: Bachelors' wives and maidens' children are well taught.

REFERENCES

ICH E6 Guideline: 1996. *Good Clinical Practice Step 4.*

Biostatistical Methodology in Clinical Trials in Applications for Marketing Authorisations for Medicinal Products. Final CPMP Guidelines III/3630/92-EN. 15 December 1994.

ICH. 1995. *Guideline on the Structure and Content of Clinical Study Reports*, Draft 13. Consensus text accepted by the ICH Steering Committee under Step 2 of the ICH process.

26

How to Handle
Suspected Fraud

Herman Pieterse
Profess® Medical Consultancy BV
Heerhugowaard, The Netherlands

Scientific misconduct is the execution of a study in a way that compromises the validity or reliability of the findings, or violates the rights of individuals who participate in the study. It covers a wide range of problems, ranging from negligence to fraud. The slippery slope from negligence to fraud is exemplified by the following situations found at investigational sites.

- *Minor negligence*: Sloppy record-keeping; disorganised data; study co-ordinator and monitor can correct the documentation.

- *Negligence*: Only some data are compromised and must be removed from the database.

- *Gross negligence*: A reckless failure to exercise adequate quality control; all patient data is compromised, and efficacy data must be removed from the database.

- *Fraud*: Deliberate actions taken to mislead the sponsor; the principal investigator should be reported to the Independent Ethics Committee or Institutional Review Board (IRB), supervisor/manager of the investigator, professional associations of the investigators, and the authorities (ICH GCP 5.20.2).

In order to constitute fraud, the following conditions must be present.

1. The investigator made a representation by words or conduct (e.g., he/she signed the Case Report Form [CRF] that contains data for which he/she is responsible).

2. The investigator made the representation, either knowing it to be untrue or being reckless as to its truth. (This can be difficult to determine; it requires either a confession by the investigator or a conclusion of wrongdoing by the auditor who investigates the case.)

3. The representation was made with the intention that the sponsor should act on it (e.g., one can assume that if the CRF was submitted to the sponsor, then the intention was that the sponsor would use the compromised data).

4. The sponsor used the data and suffered damage by doing so. ('Damage' could be a delay in product approval as a result of detection of the fraud during an inspection by the regulatory authorities. If the compromise had been detected by an auditor and the situation corrected before significant damage resulted [e.g., delays], the only recoverable damage might be payments made to the clinical trial site by the sponsor).

However, one should bear in mind that an attempt to deceive is not actionable; there must be damage to the sponsor. The relationship between an investigator and a sponsor almost always involves a contract. Scientific misconduct and data compromise are breaches of contract.

General fraud differs from errors and poor quality through intent. It is evident that inaccurate re-copying of data is a transcription error, whereas non-compliance with procedures for collecting data does not constitute fraud but rather poor quality. In clinical research, poor quality is much more frequent than fraud and, thereby, much more damaging to studies.

Errors can easily be corrected by reference to a document bearing the correct value or information. The poor quality of an investigator is more serious and should be taken into account from the moment it is detected, because it may lead to fraud if it is not handled appropriately.

Fraud is a voluntary act which implies, in general, an intention to mislead for personal gain. Fraud is not negligence or fault.

TYPES OF FRAUD

There are three types of fraud.

1. *Falsification of data.* The data actually exist, but the investigator or a member of his/her staff records false data on the CRF with the objective of making the patient eligible for inclusion in the study by avoiding a 'wrongful inclusion' or major protocol deviations. Example:

   - modification of the age of the patient based on the date of birth;

   - failure to record a pre-existing antecedent or treatment; or

   - recording incorrect dates.

2. *Invention of data.* The data do not exist with regard to the patient; data is invented or made-up. The aim is to avoid having too much missing data, which would exclude a particular patient. The 'invented' data may exist elsewhere, but are not consistent with other data from this patient. For example,

   - insertion of data between two other pieces of data (i.e., filling in of empty spaces on a CRF or on a patient diary card);

- attribution to a certain patient of data concerning another patient (e.g., ECG, laboratory value, score, etc.); or

- invention of data or even complete fabrication of a CRF.

3. *Absence or falsification of informed consent.* The objective of this activity is to keep up or increase the recruitment of patients. For example,

- the absence of consent; or

- the creation of a false consent.

WHO ARE THE PERPETRATORS OF FRAUD?

The patient can tell lies to the investigator. This is not considered to be fraud because these patients are under no legal obligation to tell the truth. The fact that the patient can conceal facts could have an effect on his/her own safety, but it only concerns the patient himself/herself.

The investigator is the one who can commit fraud. To put it in another way, fraud is most often committed under the investigator's responsibility.

HOW CAN FRAUD BE DETECTED?

Fraud can be detected by direct proof, by indirect proof, and by presumed proof.

Direct Proof

Direct proof is by far the easiest method of detecting fraud but also the rarest. The study monitor has direct access to both the correct information and the incorrect information, thereby allowing proper cross-checking. Examples of direct proof include

- information in the medical file of the patient contradicts the CRF entries;

- the handwriting or signature of the patient in a patient diary totally differs from the handwriting on the informed consent (name of patient, signature, and date);

- administrative forms, such as payments for hospital consultations, have dates completely different from the dates in the CRF; and

- photocopies which have been provided are different from the original file.

Indirect Proof

Indirect proof is much more common and requires careful planning by the perpetrator of the fraud and it is not possible to compare data from independent sources, and detection is nearly impossible. Examples of indirect proof include

- the number of results of examinations do not follow a normal pattern of progression, contrary to what appears on documents derived from the source of information (e.g., laboratory tests);

- the creation of a false set of test results from other results in data belonging to other known patients;

- alterations of copies or original documents by changing their placement; the alteration is visible, but there is only a single source document;

- the handwriting of the patient in a study document changes;

- handwriting remains the same throughout, whereas there should be as many different varieties as there are persons generating information; and

- appearance of evaluation data, when these data did not exist previously.

Presumed Proof

The study monitor is not confronted with false or incorrect data, and the realisation of fraud is not immediate. The fraud is discerned during systematic checks, or those made on demand, of clinical experimental data by the biostatistician. This situation can occur during the creation of "interpolated" or "extrapolated" data. The data transcribed may seem plausible from a statistical point of view. To detect these kinds of anomalies, several data are needed, such as

- interpolation of patient diary cards by the investigator;

- the use of statistical techniques that can result in finding the repeated use of an algorithm by the investigator; or

- the creation of multiple data or the invention of cases by the investigator.

WHAT QUESTIONS SHOULD A MONITOR ASK WHEN SUSPECTING FRAUD?

The monitor should investigate the situation by asking the following questions.

1. Which actions should I take according to the Standard Operating Procedure (SOP)?

    The SOP should indicate which steps the monitor should take, how the fraud should be identified, who has to be informed, within which period of time the issue should be resolved, and which follow-up activities should be initiated (e.g., reporting to the authorities, the execution of a for cause audit).

2. Is this case a question of fraud or a quality defect?

    A quality defect can never be repaired, whereas fraud can be repaired by excluding the data from the suspect centre. Poor quality could be caused by an incompetent investigational team inappropriately instructed or miscalculating the time and resources needed to execute a proper GCP-compliant clinical study. The procedures used in the study could be the reason for the poor quality (i.e., the procedures do not comply with the medical practice). Fraud, however, may be attributed to chance.

3. What is the impact of fraud on the study?

   If there is no or negligible impact from the fraud, the sponsor will have the tendency to look the other way and close his/her eyes to the possible complications.

4. Is it possible to easily reduce the impact of fraud in this study?

   The impact of fraud can be reduced by closing down the centre during the experimental phase of the study. When the study has been completed and the data are excluded, then the exclusion can result in suspicion as to the reasons for the closing down. When fraud is detected during the statistical analysis and the data are excluded, then suspicion regarding the manipulation of data is most likely. Accurate documentation of each event and step is recommended.

5. How can casting doubt on the company be avoided while assuring that fraud has nevertheless been taken into account?

   The industry always wishes to avoid the consequences of fraud. It wishes to be able to remove the fraudulent elements from the analysis and not be criticised for doing so. The industry must be able to identify those who have committed fraud in order to avoid using them further as investigators, and to minimise the impact on sales and the image of the company.

WHAT ARE THE CONSEQUENCES OF FRAUD?

The consequences can vary according to their importance in the clinical study, varying from the hardly perceptible change in a variable to the creation of a nonstatistically detectable bias which can render the conclusions of the study useless. This can eliminate the credibility of the entire investigating team. The discovery of fraud is tremendously expensive and compromises the sponsor once the registration or design dossier is submitted to the authorities. It is beyond any doubt that fraud should always be prevented.

HOW SHOULD THE SPONSOR REACT IN CASE OF FRAUD?

It is essential for the company to safeguard itself and to respect the rights of the investigator who is suspected of fraud. It is an absolute requirement that an adequate procedure exist in such a highly sensitive and often neurotic situation in order to reassure and assist those persons who have been accused of fraud, thus avoiding panic.

As mentioned previously, the procedure should address the following points.

- Responsibility for the management of fraud should always be taken by senior management.

- The Quality Assurance department should take the lead in compiling the dossier containing all of the information needed to make a decision. The study monitor is the liaison but not the final person responsible for the activities to be performed.

- If there are proper grounds for the suspicion of fraud, then an external, on-site audit should be arranged, because such an audit attests the elements underlying any decisions for further actions.

- Criteria should be established with respect to the follow-up activities, depending on the severity. These criteria should be general and not specific to the study in question.

The dossier generated by the Quality Assurance department should not be left alone in the face of fraud; the risk of bias of evaluation becomes realistic. The company should consider its image with regard to prescribers, the social situation of the investigator, and the possibility of not succeeding in the search for concrete proof. The company could risk accusing an innocent investigator—with unpredictable consequences. A suspicion or a body of arguments without any real proof requires that a judgement be brought based on authentic conviction. It is therefore preferred that the evaluation of the dossier be entrusted to an external organization—an independent third party.

Although the ideal situation would be one in which the authorities take these actions and perform the "for cause" audit, it is well known that these authorities cannot act fast enough. During the investigation of the fraud, the interests of the investigator and the company need to be respected.

Therefore, it is recommended that a committee be appointed according to the U.K. model (comprised of representatives of the medical profession and the industry) or the Danish model (comprised of people with a reputation for honesty and moral behaviour) to deliberate independently on such fraud dossiers. The problem needs to be treated rapidly (i.e., in less than six months), so that the company will know how to manage the data.

In summary, the sponsor should ensure that the following actions are taken.

1. The monitor immediately reports the suspicion for fraud to the manager responsible for the study and the director of the medical department. If, after studying the case, it is decided that insufficient evidence exists to prove fraud, then the data and considerations will be described in a detailed report.

2. The sponsor stops payments to the investigator.

3. An audit is scheduled to verify the data.

4. The results of the audit, described in the audit report, refer to the findings and represent either of the following situations:

   - the suspected fraud could not be confirmed by the audit. Then the medical director will determine which further steps have to be taken; or

   - the suspected fraud is confirmed by the audit, or more facts have been collected to substantiate the accusation; this is then formally reported to the medical director.

5. The management of the sponsor verbally informs the authorities. Then the authorities should consider performing a 'for cause audit'. The FDA's Division of Scientific Investigations will, according to the Compliance Program 7348.811, perform such a "for cause" audit. If deficiencies are found by the FDA inspectors which indicate that the investigator has repeatedly or deliberately violated FDA regulations, or has submitted false information to the sponsor in a required report, then the FDA will initiate actions which may ultimately result in the determination that the clinical investigator will not receive investigational devices in the future. This process has become known as the disqualification process. With respect to Europe, no such regulations currently exist for medical devices. However, it is well possible such a requirement could emerge in the near future.

6. The sponsor should decide if and what legal steps will be taken.

REFERENCES

Bogaievsky, Y. 1995. 'Fraud and Misconduct in Clinical Research: Company Dilemmas and Solutions.' *Drug Information Journal* 29:1269–1273.

Brackman, F. 1994. 'The Consequences of Fraud in Clinical Studies.' *GCP Journal* 1 (3):26–27.

Brock. P. 1994. 'Suspected Fraud in Clinical Research: Handling the Problem.' *GCP Journal* 1 (3):8–14.

Mackintosh, D.R., and V. J. Zepp. 1996. 'Detection of Negligence, Fraud, and Other Bad Faith Efforts During Filed Auditing of Clinical Trial Sites.' *Drug Information Journal* 30 (3):645–655.

'Monitors did you know?' 1995. *GCP Journal* 2 (1):27.

Schmidt, J., H. Gertzen, K. M. Aschenbrenner, and S. Ryholt-Jensen. 1995. 'Detecting Fraud Using Auditing and Biometrical Methods.' *Applied Clinical Trials* 4 (5):40–49.

Wells, F. 1994. 'The Importance of GCP in the Prevention and Detection of Fraud.' *GCP Journal* 1 (3):4–7.

Glossary

adverse event (AE) Any undesirable clinical experience that occurs with a device that results in a death, malfunction, or is life-threatening to a subject or user; any unanticipated and/or unexpected occurrence that falls outside the risk analysis of the device.

alternative hypothesis The opposite of a null hypothesis. *See also* null hypothesis, research hypothesis.

ANOVA (statistics) Analysis of variance.

audit (of a clinical trial) A systematic and independent examination of raw data and associated records to verify that the data reported are consistent with records at the investigation site and to determine whether the trial was conducted in compliance with its protocol.

audit certificate Document that certifies that a successful audit has been completed (at an investigative site, contract research organisation, or clinical research department of a device company).

audit trail The documentation of events at each step of a clinical trial process; used by auditors to trace the source and determine the authenticity of clinical trial data.

authorised representative Under the Medical Device Directives, any legal or natural person established in the Europen Union who is designated to act as the manufacturer's regulatory representative toward regulatory authorities with regard to the manufacturer's obligations under the directives.

balanced study   Trial in which a particular type of patient is equally represented in each study group.

baseline assessment   The assessment of subjects as they enter a trial and before they receive any treatment.

beta error   (statistics) *See* type 2 error.

between-patient variation   In a parallel trial design, differences between patients used to assess treatment differences.

biostatistics   A branch of statistics applied to the analysis of biological phenomena.

blind study   A study in which the patient or the investigator (or both) are unaware of what trial product a patient is taking. *See also* single-blind study.

brochure   A collection of relevant information known prior to the onset of a clinical investigation. *Synonyms:* clinical investigation brochure, Investigator's Brochure.

case record form   *See* case report form.

case report form (CRF)   (1) A standard record of data and other information about each subject in a trial. Records must be on paper or magnetic media (such as optical disks). (2) A set of documents designed for recording all relevant patient- and device-related data. It shall be produced in a way that guarantees controlled document numbering and the subject's anonymity.

categorical data   Data evaluated by sorting values into various categories (e.g., severe, moderate, and mild).

causality   Determining whether there is a reasonable possibility that the device caused or contributed to an adverse event. It includes assessing temporal relationships, dechallenge/rechallenge information, association (or lack of association) with underlying disease, and the presence (or absence) of a more likely cause.

clean database (or file)   A database in which errors have been eliminated and in which measurements and other values are provided in the same units.

clinical co-ordinator   *See* clinical research co-ordinator (CRC).

clinical investigation   *See* clinical trial.

clinical investigation brochure   *See* Investigator's Brochure.

clinical investigation plan (CIP)   A document that states the rationale, objectives, statistical design, and methodology of the trial and the conditions under which it is to be performed and managed. *Synonyms:* study protocol.

clinical investigation project plan (CIPP)   Principal document including detailed information on how the clinical investigation is to be implemented and closed out.

clinical research co-ordinator (CRC)   The person who handles most of the administrative responsibilities of a clinical trial. He/she acts as a liaison between the investigative site and the sponsor, and reviews all data and records before the monitor's visit. *Synonyms:* trial co-ordinator, study co-ordinator, research co-ordinator, clinical co-ordinator.

clinical significance   The change in a subject's clinical condition regarded as important, whether or not due to the test article. Some statistically significant changes (in blood tests, for example) have no clinical significance. The criterion or criteria for clinical significance should be stated in the protocol.

clinical study report   A comprehensive and complete description of the results of an investigation after its completion, including a description of methodology and design, data analysis together with a critical evaluation and any statistical analysis. The clinical study report must be signed by the sponsor and investigators.

clinical trial   (1) Systematic study of a test article (e.g., treatment, drug, device) in one or more human subjects. *Synonyms:* clinical study, clinical investigation (21 CFR 50.3). (2) Any systematic study in subjects undertaken to verify the performance of a specific device under normal conditions of use (synonym: clinical investigation).

clinical trial materials   The complete set of supplies provided to an investigator by the trial sponsor.

coding   In clinical trials, the process of assigning data to categories for analysis. Adverse events, for example, may be coded using the COSTART system (Coding dictionary: *Coding Symbols for Thesaurus of Adverse Reaction Types*).

coding dictionaries   List of associated codes with text on a common theme (e.g., adverse events). Each term has a unique reference.

cohort   A group of subjects in a clinical trial followed-up at regular, pre-determined intervals.

cohort study   *See* prospective study.

comparative study   A study in which the investigative device is compared against another product.

Competent Authority (CA)   The regulatory body, at European member state level, responsible for monitoring compliance with the regulations.

confidentiality (of records)   Protecting the privacy of trial subjects, including their identity and all personal medical information. Data verification procedures, when required, may be carried out only by a properly authorised person, keeping identifiable personal details in strict confidence. Consent to the use of records for data verification should be included in the informed consent procedure.

Conformity Assessment   The process by which compliance with the Euopean Directives is carried out.

contract research organisation (CRO) Company or institution that contracts to conduct research tasks and assumes research responsibilities for a sponsor.

controlled study A study in which a test article is compared with a treatment that has known effects. The control group may receive no treatment, a standard treatment, or a placebo.

co-ordinating centre The headquarters for a multi-site trial that collects all data.

correlation The relationship of one variable to another, not to be confused with causation.

crossover trial In crossover trials, each subject receives both treatments being compared or the treatment and control. Such trials are used for patients who have a stable, usually chronic, condition during both treatment periods.

curriculum vitae (CV) A document that outlines a person's educational and professional history.

database Data stored in computerised form for retrieval, processing, and/or analysis.

data monitoring A process by which case report forms are examined for completeness, consistency, and accuracy.

Declaration of Helsinki A set of recommendations or basic principles that guide medical doctors in the conduct of biomedical research involving human subjects. It was adopted by the 18th World Medical Assembly (Helsinki, Finland, 1964) and revised by the 29th (Tokyo, Japan, 1975) and 35th (Venice, Italy, 1983) World Medical Assemblies. The full text is available in various reference materials and on the Internet.

demographic data The characteristics of subjects or study populations, which include information such as age, sex, family history of the disease or condition for which they are being treated, and other characteristics relevant to the study in which they are participating.

documentation   All records in any form (including those on paper and on magnetic and optical storage media) that describe trial methods and conduct, including the study protocol, copies of regulatory submissions and approvals, copies of Institutional Review Board/Ethics Committee submissions and approvals, investigator curricula vitae, consent forms, monitor reports, audit certificates, relevant correspondence, laboratory reference ranges, raw data, completed case report forms, and the final report.

effectiveness   The desired measure of a drug's influence on a disease condition as proved by substantial evidence from adequate and well-controlled investigations.

efficacy   A product's ability to produce beneficial effects on the course or duration of a disease.

end point   An indicator measured in a subject or biological sample to assess the safety, efficacy, or other objective of a trial. *See also* surrogate marker.

Essential Requirements (ERs)   The requirements laid down by the directives in Annex I and by which compliance must be demonstrated as part of the CE marking process.

Ethics Committee   *See* Institutional Review Board.

exclusion criteria   A list of criteria, any one of which excludes a potential subject from participation in a study.

external consistency   The consistency of a procedure between sets of data.

first-study-in-man study   The first study in which the test product is administered to human beings.

Food and Drug Administration (FDA)   The U.S. regulatory authority charged with, among other responsibilities, granting investigational device exemptions and pre-market approvals.

Good Clinical Practice (GCP)   The standard for the design, implementation, conduct, and reporting of clinical trials to the public that the data are credible and that the rights of subjects are protected.

harmonised standard A European Norm that supports the directive's essential requirements and has been published in the *Official Journal of the European Community.*

human subject A human subject, or individual who is or becomes a participant in research; subject may be either a healthy human or a patient. *Synonym:* subject.

Huriet Law France's regulations covering the initiation and conduct of clinical trials.

IDE Investigational device exemption (U.S. Food and Drug Administration).

inclusion criteria The criteria which prospective subjects must meet to be eligible for participation in a study.

informed consent (1) The voluntary confirmation of a subject's agreement to take part in a particular trial, documented in accordance with the relevant regulations of the legal jurisdiction in which the trial takes place. Consent should be sought only after subjects have been given full information about the trial (e.g., explanation of its objectives, potential benefits, risks and inconveniences, and the subject's rights and responsibilities as described in the Declaration of Helsinki). Requests for consent should take place under circumstances that minimise the possibility of undue influence and that provide prospective subjects and representatives sufficient opportunity to consider potential risks and benefits explained in a language they can understand. (2) The process documentation of obtaining confirmation of the willingness of a subject or legal guardian or representative to participate in an investigation after all relevant information on the trial has been given to the subject.

inspection An official audit to verify adherence to Good Clinical Practice, conducted by relevant authorities at an investigation site and/or at the sponsor.

Institutional Review Board (IRB) (1) A board, committee, or other group formally designated by an institution to review biomedical research involving human subjects. In the United States, an Institutional Review Board reviews the protocol and approves initiation of and periodically reviews the conduct of such research. All clinical studies performed in the United States, as well as those conducted in many other countries, require the approval of—or notification to—an Institutional Review Board or similar body. *Synonyms:* Board, Ethics Committee. (2) Independent and properly constituted body of medical professionals and non-medical members whose responsibility is to ensure that the safety, well-being, and human rights of the subjects participating in a proposed investigation are protected.

internal consistency   The consistency of a procedure within a set of data.

investigator   (1) The person(s) responsible for conducting a trial and for the health and welfare of subjects during the investigation. Multi-centre trials may have a co-ordinating, or principal, investigator; (2) the individual under whose immediate direction a test article is administered, or dispensed to, or used on a subject, or the responsible leader (principal investigator) of a team of investigators. *See also* sponsor-investigator.

Investigator's Brochure   A document provided to investigators by the sponsor of a clinical trial that contains a collection of information about the development of a device. It provides information on the technical development, in-vitro and ex-vivo performance testing, biocompatability, and safety testing in animals, and the clinical experience. It also summarises any safety and effectiveness data from prior studies and, on the basis of prior studies, describes potential risks and side effects. It also recommends precautions or special monitoring appropriate to the investigational use of the product under study. When new data emerge during the course of the trial, information in the brochure is updated.

longitudinal study   An investigation in which data are collected from a number of subjects over a long period of time. (A well-known example is the Framingham Study, a study on the incidence of hypertension in more than 5,000 men and women over a long period.)

matched pair design A type of parallel trial design in which investigators identify pairs of subjects who are identical with respect to relevant factors, then randomise them so that one receives treatment A and the other treatment B. *See also* pairing.

mean The sum of the values of all observations or data points divided by the number of observations; an arithmetical average.

median The middle value in a data set, that is, just as many values are greater than the median as are less than the median.

medical device Any instrument, apparatus, appliance, material, or other article, whether used alone or in combinations, including software necessary for its proper application, that is intended by the manufacturer to be used for human beings for the purpose of

- diagnosing, preventing, monitoring, and treating diseases;

- diagnosing, preventing, monitoring, treating, alleviating, or compensating for an injury or handicap;

- investigation, replacement, or modification of the anatomy or a physiological process; or

- the control of contraception,

and which does not achieve its principal intended action in, at, or on the human body by pharmacological, immunological, or metabolic means, but which may be assured in its function in regard to such means.

medical practice computer system A PC- or network-based computer system used to manage electronic subject files. Such a system is neither sponsor-supplied nor trial specific.

megatrials Massive, randomised clinical trials that test the advantages of marginally effective experimental drugs by enrolling 10,000 or more subjects. *Synonym:* large-sample trials.

Memorandum of Understanding (MOU) A statement reached between various states or their regulatory agencies to allow the mutual recognition of regulatory activities.

meta-analysis A statistical process for pooling data from many clinical trials to give a clear answer.

mode The most frequently occurring value in a data set.

monitor (1) A person employed by the sponsor or contract research organisation who is responsible for determining that a trial is being conducted in accordance with the protocol. A monitor's duties may include, but are not limited to, helping to plan and initiate a trial; assessing the conduct of the trial; and assisting in data analysis, interpretation, and extrapolation. Monitors work with the clinical research co-ordinator to check all data and documentation from the trial. (2) A person appointed by the sponsor and responsible to him/her for monitoring and reporting on the progress of the investigation.

multi-centre study A trial conducted under a single protocol but at more than one investigational site and by more than one investigator. Special issues in multi-centre trials include the need to balance treatments at each centre and to ensure consistency in assessment methods and measurements across all centres.

Notified Body (NB) An organisation appointed by the Competent Authority and assessed to compliance by an accreditation body, who is qualified to verify compliance with the requirements of the device directives.

null hypothesis An hypothesis (e.g., subjects will experience no change in blood pressure as a result of administration of the test product) used to rule out every possibility except the one the researcher is trying to prove; an assumption about a research population that may or may not be rejected as a result of testing. It is used because most statistical methods are less able to prove something true than to provide strong evidence that it is false. *See also* research hypothesis.

Nuremberg Code A code of ethics for conducting medical research set forth in 1947.

objective measurement A measurement that cannot be influenced by investigator bias; for example, blood glucose levels or electrocardiogram tracings.

open-label study   *See* open study.

open study   A trial in which subjects and investigators know which product each subject is receiving; opposite of double-blind study.

pairing   A method by which subjects are selected so that two subjects with similar characteristics (e.g., weight, smoking habits) are assigned to a set; one receives treatment A and the other receives treatment B.

parallel design trial   *See* parallel trial.

parallel group trial   *See* parallel trial.

parallel trial   A trial in which subjects are randomised to one of two differing treatment groups (usually medicine and placebo) and usually receive the assigned treatment during the entire trial. *Synomyms:* parallel group trial, parallel design trial.

patient   A person under a physician's care for a particular disease or condition. *See also* human subject.

patient file   A file containing demographic, medical, and treatment information about a patient or subject. It may be paper based or a mixture of computer and paper records.

period effect   A designated period during the course of a trial in which subjects are observed and no treatment is administered.

PMA   Pre-market approval (U.S. Food and Drug Administration).

post-marketing surveillance   A mechanism to monitor device performance after commercialisation. This includes analysis and correlation of all information pertaining to a device such as literature, scientific publications, complaints, and marketing information.

pre-clinical studies   Technical testing, including animal studies, of a device against standards and essential requirements, whereby data establishes safety, a device's performance, and the boundaries for safe use of the device for subsequent human testing (i.e., in clinical studies or trials).

principal investigator   The investigator appointed to co-ordinate the work in an investigation or of several investigators at one site.

prospective study   An investigation in which a group of subjects is recruited and monitored in accordance with criteria described in a protocol.

p value   (statistics) The lowest level of significance at which a given null hypothesis can be rejected, that is, the necessary criterion for determining that the result probably did not happen by chance. *See also* statistical significance.

qualitative variable   A variable that cannot be measured (happiness, for example).

quantitative variable   A variable that can be measured (blood pressure, for example).

random allocation   The assignment of subjects to treatment (or control) groups in an unpredictable way. Assignment sequences are concealed, but available for disclosure in the event a subject has an adverse reaction.

randomisation   A process that aims to prevent bias by secretly and arbitrarily assigning subjects to treatment or control groups.

random number table   A table of numbers with no apparent pattern used in the selection of random samples for clinical trials.

random sample   A population of which the members are selected by a method designed to make it probable that each person in the target group has an equal chance of selection.

raw data   Records of original observations, measurements, and activities (such as laboratory notes, evaluations, data recorded by automated instruments), without conclusions or interpretations.

recruitment (investigators)   The process used by sponsors to select investigators for a clinical study.

recruitment (subjects)   The process that employs inclusion and exclusion criteria and is used by investigators to enroll appropriate subjects into a clinical study.

recruitment period   The time-period during which investigators must complete enrollment of their quota of subjects for a trial.

recruitment target   The number of subjects that must be recruited into a study to meet the requirements of the study protocol. In multi-centre studies, each investigator has a recruitment target.

representative   An individual or judicial or other body authorised under applicable law to consent on behalf of a prospective subject to that subject's participation in the procedure(s) involved in the research.

research co-ordinator   *See* clinical research co-ordinator (CRC).

research hypothesis   The conclusion a study sets out to support (or disprove). For example, blood pressure will be lowered by [specific endpoint] in subjects who receive the test product. *See also* null hypothesis.

risk   The probability of harm or discomfort a device could pose for subjects or patients. Risks are either acceptable or unacceptable. They are analysed by using various risk management tools such as Failure Modes and Effects Analysis (FMEA) and Fault Tree Analysis (FTA), for example.

safety   Relative freedom from harm. In clinical trials, safety refers to an absence of harmful side effects resulting from the use of the product and may be assessed by laboratory testing of biological samples, special tests and procedures, psychiatric evaluation, and/or physical examination of the subjects.

single-blind study   A study in which subjects do not know whether they are receiving the active drug or a placebo.

source data   Original records from which forms, such as case report forms, are completed. Examples include patient files, electrocardiograms, electroencephalograms, X-ray films, and laboratory notes.

Source data validation (SDV)   Checking of data in the CRFs and the trial against original source documents.

sponsor (1) A person or organisation that initiates and/or finances a clinical trial but does not actually conduct the investigation. (2) An organisation of individuals that takes responsibility for the initiation and implementation of clinical investigations.

sponsor-investigator One who both initiates and actually conducts (alone or with others) a clinical investigation; it does not apply to corporations or agencies.

standard deviation (statistics) An indicator of the relative variability within a group; the square root of the variance.

standard operating procedure (SOP) Written procedures controlled in the framework of quality system or under GCP, that detail the process of a clinical trial. SOPs provide a general framework to provide the means of efficient implementation and performance of all the functions and activities for the trial.

statistical significance The level at which an investigator can conclude that observed differences are not due to chance alone; for example, a p value of 0.05 (also called significance at the 0.05 level) indicates that there is about 1 chance in 20 that the differences observed occurred by chance alone.

stochastic Involving a random variable; involving chance or probability.

surrogate marker A measurement of a drug's biological activity that substitutes for a clinical end point, such as death or pain relief.

t-test A statistical test used to compare the means of two groups of test data.

type 1 error (statistics) An error made when a correct null hypothesis is rejected.

type 2 error (statistics) An error made when an incorrect null hypothesis is not rejected.

unequal randomisation   A technique used to allocate subjects into groups at a differential rate. For example, three subjects may be assigned to a treatment group for every one assigned to the control group.

validation of data   A procedure carried out to ensure that the data contained in the final clinical trial report match original observations.

validity   The accuracy of the relationship between two or more variables.

variance   A measure of the amount by which a value differs from the mean.

weighting   An adjustment in a value on the basis of a judgement by the investigator.

within-patient differences   In a crossover trial, variability in each patient that is used to assess treatment differences.

Index